T0067551

ENGINEERING CHEMISTRY

ENGINEERING CHEMISTRY

RARE TOPICS

S. EKAMBARAM

PARTRIDGE
A Penguin Random House Company

ISBN:	Softcover	978-1-4828-6974-3
	eBook	978-1-4828-6973-6

Print information available on the last page.

To order additional copies of this book, contact
Partridge India
000 800 10062 62
orders.india@partridgepublishing.com

www.partridgepublishing.com/india

Contents

Dedicated to
Mr. R. Shanmugam, Ex. M.L.A
Perambakkam.

Preface

Engineering Chemistry with rare topics is aimed to provide readers to understand easily and quickly without having any burden on studying this book. Thus, this book starts with advanced and structural ceramics which is followed by cements, refractories and abrasives, water treatment, environmental pollution, chemistry of explosives, metallurgy, topical interest of nano materials, conventional and non-conventional energy sources, spectroscopy and ends with dyes and pigments.

I am obligated to acknowledge my wife, S. Kalyani and my daughter, E. Nandhini for their constant and continuous encouragement and support from out of their ways to write this book. I acknowledge my parents, father-in-law, relatives and friends for their expectations to be partly fulfilled from writing this book.

S. EKAMBARAM.

Advanced and Structural Ceramics

Objectives:

1. To define ceramics, types of ceramics and their applications
2. To differentiate properties of advanced ceramics from engineering plastics and metals
3. To classify the advanced ceramics based on their properties.
4. To classify the ceramics based upon their solid state structures
5. To classify the advanced ceramics based upon their applications.
6. To define advanced ceramics
7. To elaborate atomic level bonding in advanced ceramics.
8. To define and elaborate the point defects such as Schottky and Frenkel defects
9. To outline the making of point defects
10. To highlight the use of point defects.
11. To outline the mechanical properties of ceramics.
12. To highlight the electrical properties of ceramics.
13. To elaborate the relationship between microstructure and electrical conductivity of ceramics.
14. To introduce High-Temperature Ceramic Oxide (HTC) superconductors

15. To outline the natures of HTC superconductors.
16. To highlight the crystal chemistry of cuprate superconductors.
17. To briefly outline the theory of HTC cuprate superconductors
18. To introduce Meisner effect.
19. To highlight applications of HTC superconductors
20. To outline Ferromagnetic semiconductor ceramics
21. To elaborate bioceramics including applications and types of bioceramics.
22. To outline coordination environment of hydroxyapatite
23. To introduce magnetic ceramics, types of magnetic ceramics and description of magnetic ceramics.

Introduction:

The materials can be grouped into five categories and these are

1. Metals
2. Polymers
3. Ceramics
4. Semiconductors and
5. Composites.

Among the five categories, ceramics is the subject of this chapter.

Ceramics:

Ceramics are typically characterized by combination of three types of bonding such as covalent, ionic and metallic. Covalent and ionic bondings exist generally in ceramics. Ceramics are different from complex or coordinate molecules. Thus, ceramics do not have discrete molecules rather than ceramics comprise of array of interconnected atoms. Universally, ceramics are oxides, nitrides and carbides of metals or metalloids and nonmetals. In addition, diamond and graphite are also considered as ceramics.

Thus, ceramics is defined as non-metallic, inorganic solids.

Types of ceramics and their applications:

Ceramics are classified into conventional and advanced ceramics. Conventional ceramics are universally based on clay and silica. Advanced ceramics include newer materials with various properties of laser host, piezoelectric, dielectric, mechanical etc.

Advanced ceramics refers to ceramic materials that are chiefly

(1) Highly specialized by exploiting unique properties such as electric, magnetic, optical, mechanical, biological and environmental.
(2) Well performing even under extreme conditions such as high temperature, high pressure, high strength, high radiation and high corrosive exposure
(3) Mostly inorganic, nonmetallic.

Table 1 compares the important properties of metals, engineering plastics and advanced ceramics.

Table 1: Comparison of properties of Engineering Materials

Property	Metals	Engineering Plastics	Advanced ceramics
Temperature of continuous use (°C)	1000°C(typical) 1500°C (max)	250°C(typical) 350°C(Max)	1200°C(typical) 2500°C(max)
Hardness	Medium – high	Low – medium	High
Toughness Flexibility Impact Resistance	Medium – high High	High Medium-high	Low Low
Corrosion resistance	Low – medium	Medium	High
Coefficient of thermal expansion	High	Medium	Low
Electrical Properties	Conductive	Insulator to conductive with fillers	Insulator to conductive
Density	High	Low	Medium

Ceramics are classified as shown below in the outline 1.

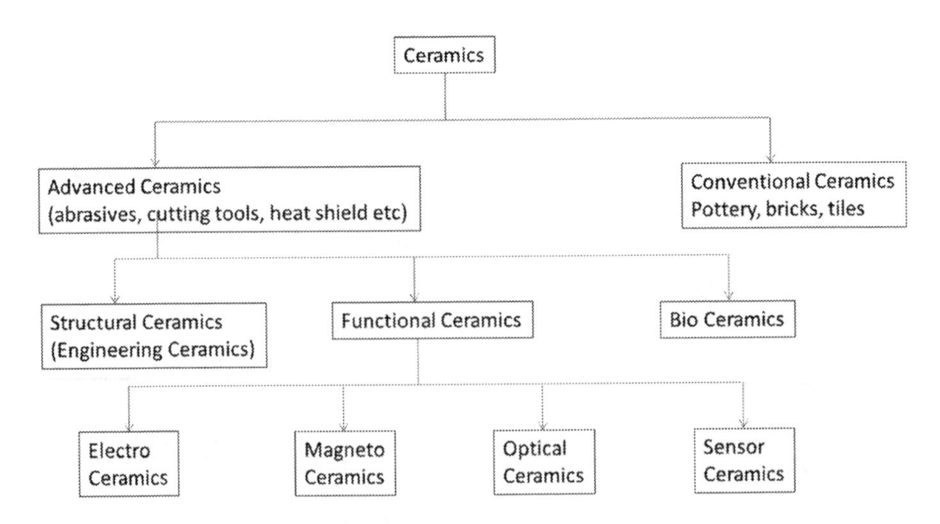

Outline 1: Classification of ceramics

Examples of Ceramics:

Some examples of ceramics that are based upon their structures are summarized below in Table 2.

Table 2: Structures and examples of ceramics

Type	Type of Bonding	Examples of Ceramics
Binary Compounds	NaCl	MgO, TiC, PbS
	GaAs	Beta-SiC
	AlN	BeO, ZnO, AlN
	CaF_2	C-ZrO_2, CeO_2, UO_2
	Al_2O_3	Corundum (Al_2O_3)
Spinel	AB_2O_4	$MgAl_2O_4$, $NiFe_2O_4$, Fe_3O_4
Perovskite	ABO_3	$CaTiO_3$
Garnet	$A_3B_5O_{12}$	$Y_3Al_5O_{12}$
Zerolite & Microporous	-	$(N_{a2},k_2,Ca,Ba)(Al,Si)O_2.xH_2O$

Table 3 below summarizes some examples of ceramics with their useful properties.

Table 3: Applications of ceramics

Application Field	Component	Useful Properties	Advanced ceramics
Processing Technology	Chemical apparatus	Corrosion resistance	Aluminum oxide
	Sliding rings	Wear resistance	Silicon carbide
	Thread guides		Titanium oxide
	Wire-drawing dies		Zirconium oxide

Shaping of materials	Cutting tools	Wear resistance	Aluminum oxide
	Grinding disks	Hardness	Silicon nitride
	Sand-blasting nozzles		Silicon carbide
High temperature	Burner nozzles	Heat resistance	Silicon nitride
	Welding nozzles	Corrosion resistance	Silicon carbide
	Heat exchanger	Heat conductivity	Aluminum oxide
	Crucibles		Carbon
	Heat pipes		Boron nitride
Engines	Valve seats	Heat resistance	Aluminum titanate
	Turbocharger	Corrosion resistance	Silicon carbide
	Gas turbine	Heat conductivity	Silicon nitride
	Catalyst support		Cordierite
Electronics, electrical engineering, optics	Substrates	Special electrical & magnetic properties	Aluminum oxide
	Capacitors		Titanate perovskites
	Sensors		Ferrites
	Laser materials		$Y_3Al_5O_{12}$:Nd
	Magnets		Ferrites
Energy technology	Nuclear fuel	Radiation resistance	Uranium oxide
	Solid-state electrolyte	Ionic conductivity	Zirconium oxide
			Beta aluminum oxide
Medical technology	Hip joints	Mechanical resistance	Aluminum oxide
	Bone replacement	Surface finish	Calcium phosphate

With this brief introduction of this chapter, the rest of the chapter is devoted to advanced ceramics.

Advanced Ceramics:

Advanced ceramics are nonmetallic, nonpolymeric materials and these are hard, resistance to heat and chemicals & can be designed to have special properties including optical, electrical, magnetic and sensor. Thus, advanced ceramics are sophisticated products. Advanced ceramics include properties that are superior to conventional ceramics such as high mechanical strength & fracture toughness, wear resistance, refractory, dielectric, magnetic and optical properties. Governing the microstructure of advanced ceramics arises from high purity synthetic powders. Also, advanced ceramics are called fine ceramics.

When value added ceramics find diverse applications with indispensable ones in modern society, brittleness of ceramics is a critical drawback. Hence, a brittle material does not deform under load. However, recent advances in ceramics have succeeded in alleviating the problem of brittleness and hence, provided greater control over aspects of composition and microstructure as well.

With these unique and great properties of ceramics, advanced ceramics have played critical roles in the development of new technologies such as computers and telecommunications and of course, they will continue to play a leading role in the technologies of the future.

Bondings in Ceramics:

A ceramic's characteristics properties stem from its structure, both at atomic level and scales ranging from micometers to millimeters.

Atomic level bonding:

Ionic and covalent bondings are encountered at the atomic level bondings of ceramics.

In ionic bonding transfer of electrons between two neighboring atoms take place. Thus, the atom that gives up electrons becomes positive charged and the atom that accepts the electrons becomes negative charged. The two opposite charge of neighboring atoms bind together to make material.

In covalent bonding electrons are shared more or less likewise between neighboring atoms. Electrostatic force of attraction between adjacent atoms in the covalent bonding is less than it is in ionic bonding. However, the hardest material known, diamond is composed of covalently bonded carbon atoms.

Both in the ionic and covalent bondings, atoms form a group, is called unit cells. The unit cells may be repeated periodically throughout the material. Such a periodically ordered array of unit cells constitutes a crystal. If there is no existence of periodically ordered array of unit cells, the material is non-crystalline. Figure 1 below illustrates the two types of ionic and covalent bondings in ceramics.

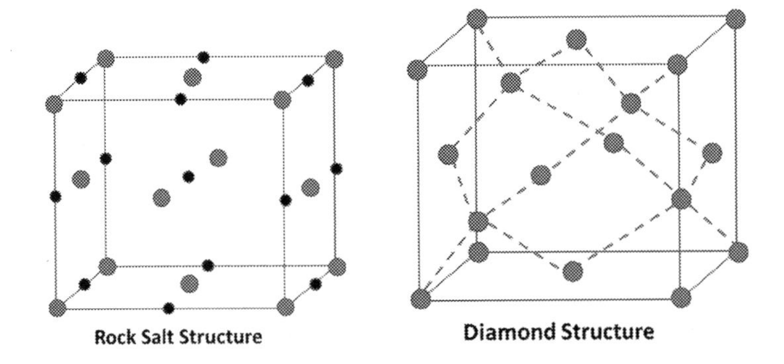

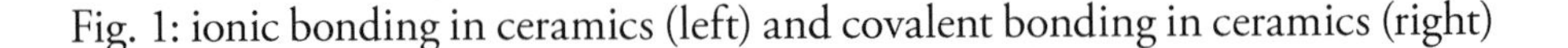

Fig. 1: ionic bonding in ceramics (left) and covalent bonding in ceramics (right)

The formation of crystalline and non-crystalline SiO_2 is explained now. If silicon dioxide/silica, SiO_2 is melted and allowed to cool gradually with controlled way, the silica molecules arrange themselves into a lattice to form crystalline SiO_2 with long range order. On the other hand, if silica melted is cooled quickly, the molecules do not have enough time to construct a crystal lattice and sudden cooling leads to frozen the irregular arrangement to produce a noncrystalline substance. Outline 2 below illustrates the effect of rapid and slow cooling of melted silica.

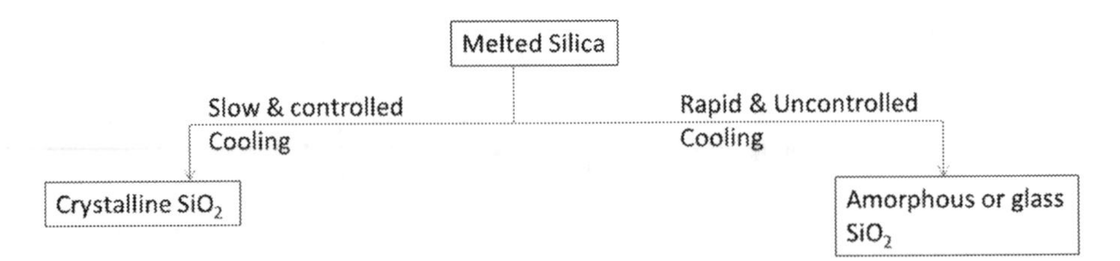

Outline 2: Effect of slow & controlled cooling and rapid & uncontrolled cooling of melted Silica.

It is absolutely clear from the above scheme that both crystalline and non-crystalline materials are composed of the same elements such as Silicon and Oxygen. But, in the case of crystalline SiO_2, the fundamental pattern of silicon and oxygen atoms is repeated habitually throughout the material. On the other hand, non-crystalline (amorphous or glass) SiO_2, there is no long-range periodicity in its atomic structure. These two structures of crystalline and non-crystalline SiO_2 are shown below in Fig.2

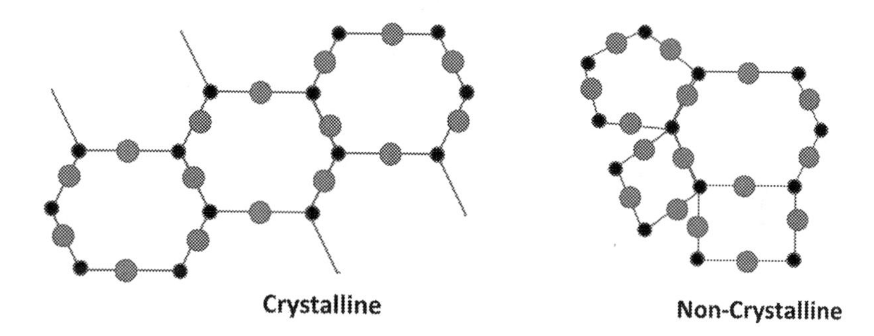

Fig.2: Crystalline and non-crystalline structural difference

Defects:

Defects are classified into three types as shown in the outline 3 below.

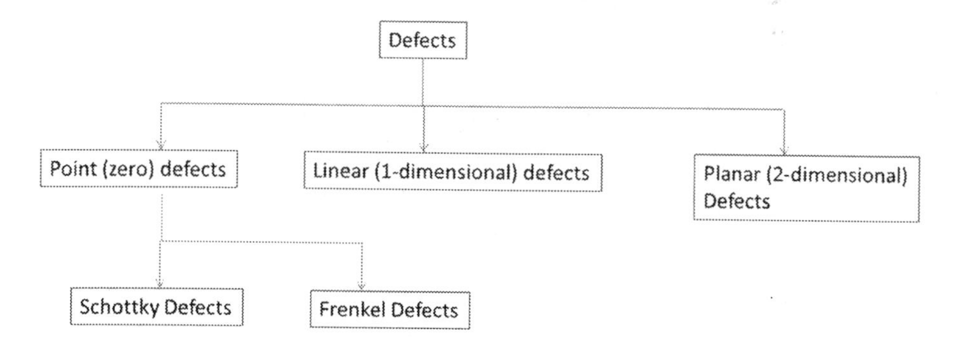

Outline 3: Classification of Defects

Point Defects:

Point defects do exist in ceramics and in fact, point defects play an important role in the determination of properties of ceramics. Therefore, it is worth to study and to understand the point defects that are present in ceramics. And how are the point defects correlated to properties of ceramics. Therefore, in this section, point defects are the subject of interest. The simplest type of defect in crystalline solid is a point defect. It is a zero dimensional defect.

Schottky Defects (due to vacancies in ionic ceramics):

If an atom is missing from the site that it should occupy in a perfect crystal, thus vacancy is created at the missing site. If several atoms are missing from the crystal, then, the site of vacancy is called schottky defects. For example, consider MgO ceramic in which, pair of vacancies are created. Thus, one on the Mg sub-lattice and another on the "O" sub-lattice are vacancies to get a pair of vacancies that witnessed in MgO. Equally, in the spinel, AB_2O_4, Schottky defects (Fig.3) consists of seven vacancies.

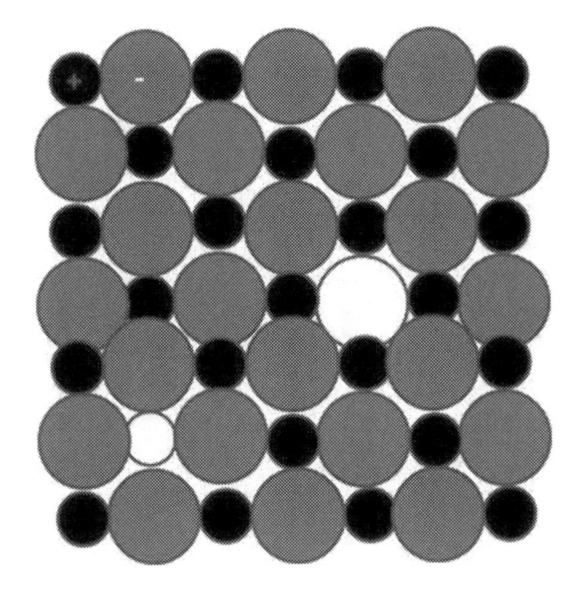

Fig. 3: Schottky Defects representation in MgO ceramic where + = Mg^{2+} and - = O^{2-}

Frenkel Point Defects (due to combination of vacancy and interstitial atom in ionic ceramics):

Frenkel defects are formed by vacancy due to removal of atom from the site where it is supposed to be and by formation interstitial atom from the atom as illustrated in the Figure 4.

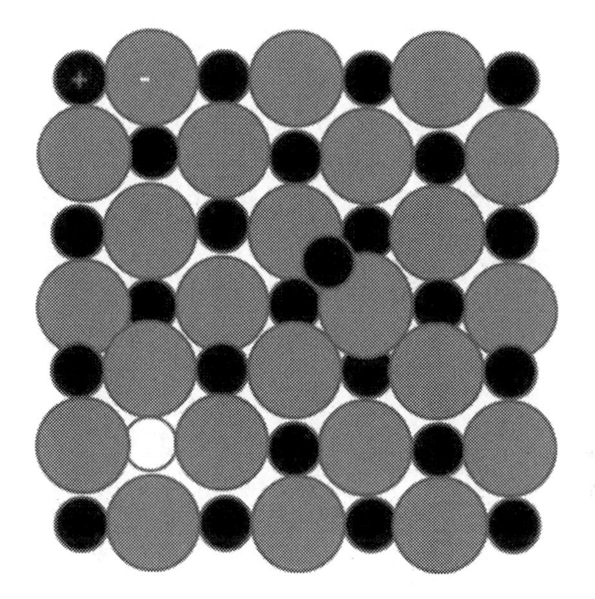

Fig. 4: Illustration of Frenkel defects formation due to combination of vacancy and interstitial atom.

Antisite Point Defects (Due to misplacement of atoms in covalent & ionic ceramics)

Yet another point defects are due to antisite defects, associated centers, solute atoms and electronic defects. The antisite defects are ordinarily perceived in simple covalent ceramics and complex ionic ceramics such as spinels ($MgAl_2O_4$) and garnets ($Y_3Al_5O_{12}$). Associated centers are formed when ceramics are exposed to ionizing radiation such as X-rays and gamma rays.

Creation of Point Defects:

Point defects can be created on their own in non-stoichiometric oxides either by aliovalent ions doping or by annealing in pure oxygen or hydrogen atmosphere.

Diffusion:

Diffusion of point defects occurs due to gradient in chemical potential. Fick's laws of diffusion are shown by the diffusion equation. Thus, the Fick's first law is given as shown below.

$$J = -D\left(\frac{dc}{dx}\right)$$

Where J = flux of diffusion species

X =direction due to concentration gradient

D = proportionality coefficient (diffusion coefficient)

Fick's second law is

$$\left(\frac{dC}{dt}\right) = D\left(\frac{d^2C}{dx^2}\right)$$

Application of Point defects:

Generally, ceramics are insulators. Therefore band gaps of them are large enough > 5 eV and hence, electronic conduction is very difficult. Under such situations, mechanism for conduction of change is commonly achieved by movement of ions. Ionic conduction responsible for conduction is called ionic conductors. Therefore, the point defects play an important role in the ionic conduction by their diffusion from higher concentration regions into lower concentration regions (due to concentration gradient).

Major mechanical properties of ceramics:

The atomic structure of ceramics engenders a chemical stability and hence, its degradation by dissolution in solvents is alleviated. Also, many ceramics are metal oxides, which are prepared by combustion at higher temperature leads to non-degradable ceramics. The strength of the bonds in ceramics also obviously yields them with a high melting point, hardness and stiffness.

Because of the same reason of strong bonding between of atoms present in ceramics it makes ceramics hard for planes of atoms from sliding simply over one another on loading. Thus, the ceramics cannot deform to relieve the stresses imposed by a load like ductile metals such as copper do.

Brittle Fracture:

Ordinarily, ceramics maintain their shape marvelously under stress until a certain threshold is exceeded. At the point of fracture threshold, the bonds unexpectedly break down and the material catastrophically fails. This is called brittle fracture. Thus, the brittle fracture is an important characteristics of ceramics and glasses. The brittle nature of ceramics is illustrated by stress – strain curves. The stress – strain relationship curve for ceramics yields tensile strength and compressive strength of any ceramics.

Consider Al_2O_3 ceramics for its tensile strength performance.

The stress – strain curve shows that the alumina ceramics survived up to 280 MPa but beyond this value, fracture of alumina ceramics is occurred. The formation of crack beyond the point of tensile strength and its propagation to end with breaking.

These two tensile and compressive strength show that ceramics have somewhat low tensile strength (280 MPa for alumina) while compressive strength is relative higher (2100 MPa for alumina) when they are compared to that of metals and alloys.

Ceramics have crystal structure such that they do not truly readily deform. As a consequences a narrow crack, which concentrates tensile stresses (arrows) at its tip to exceed the threshold at which the material's bonds are broken. This phenomenon is known as Brittle Ceramic Materials. The formation and propagation of crack to the breakage of ceramics are illustrated below in three Figures 5a, 5b and 5c.

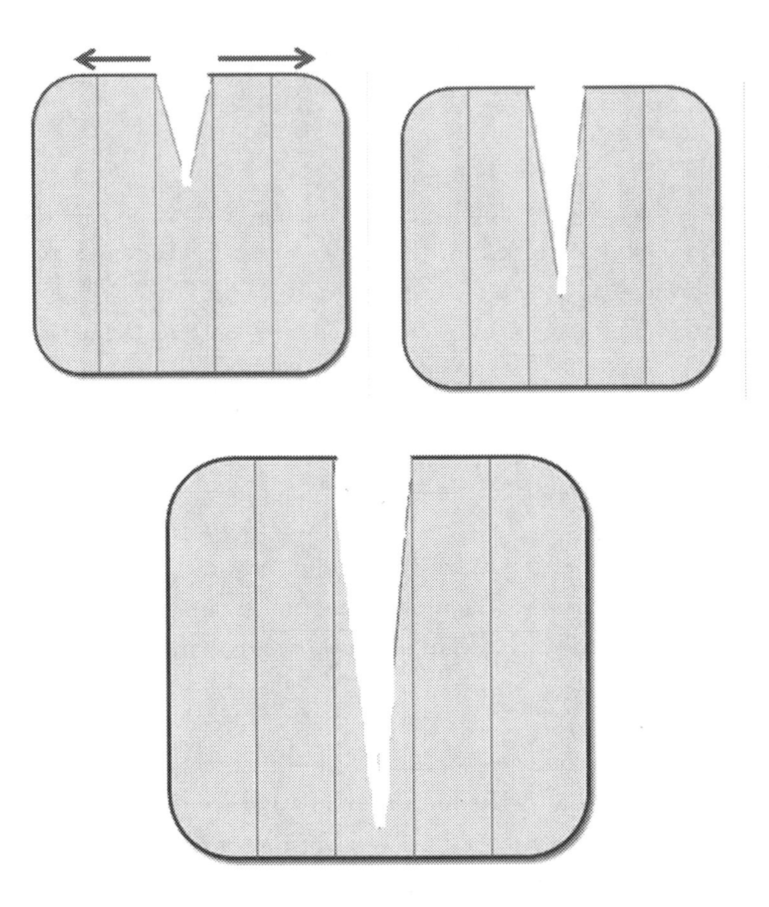

Fig. 5: Illustration of formation and propagation of tensile stress, lines are indicating the plane of atoms.

Electrical Properties of Ceramics:

Electrical properties of ceramics principally depend upon their crystal structures and atomic composition. Electrical properties of ceramics such as cobalt oxide, vanadium dioxide and rhenium trioxide are compared with copper metal. Thus, cobalt oxide, CoO forms a semiconducting band structure and hence, its behavior of electrical conductivity linearly increases from 500°K. Vanadium dioxide, VO_2 is a yet another semiconductor and hence, its electrical conductivity increases linearly up to 350°K. But, at the temperature of 350°K, its structure changes such that its electrical conductivity behaves like metal. However, electrical conductivity of rhenium trioxide is a metallic characteristic.

Effect of Microstructure on electrical conductivity of ceramics (Varistor):

Microstructure determines electrical conductivity of ceramics. However, microstructure of ceramics is dependent on atomic composition of ceramics. In this section, effect of microstructure / dopants on electrical conductivity of ceramics is explained with an example of semiconducting Zinc Oxide.

By doping impurities in the crystal structure, semiconducting ZnO can be made into conductor. By using impurity doped ZnO in combination with insulating ceramic matrix, a new type of material is made. This new type of material is called varistor. Varistor is defined as material that has variable resistor. The triumph of varistor is explained as follows. In the composite grains of insulating and varistor ZnO, habitually insulating material between varistor grains prevents current flow through the varistor at low voltages. However, electrons gain enough energy to overcome the conduction barrier between the grains, and they travel freely through the varistor at higher voltage.

In another ceramic, so called ferroelectric ceramic can be fruitfully explained by considering material atomic architecture. In the ferroelectric ceramic, uneven charge distribution due to ions in the crystal makes grains polarized. I.e. One side of grain is positive charge and another side of the grain is negative charge. However, net polarization in the ceramic grains is typically zero due to random orientation of polarized grains results in cancelling the charges each other. The unsurprisingly observed zero net polarization of grains can be made to observe polarized grains line up macroscopically. Under such polarization line up ceramics remains a nonconductor. This is the property explored in the manufacture of capacitors, which store electric charge. The function of capacitors is illustrated in the figure 6 below.

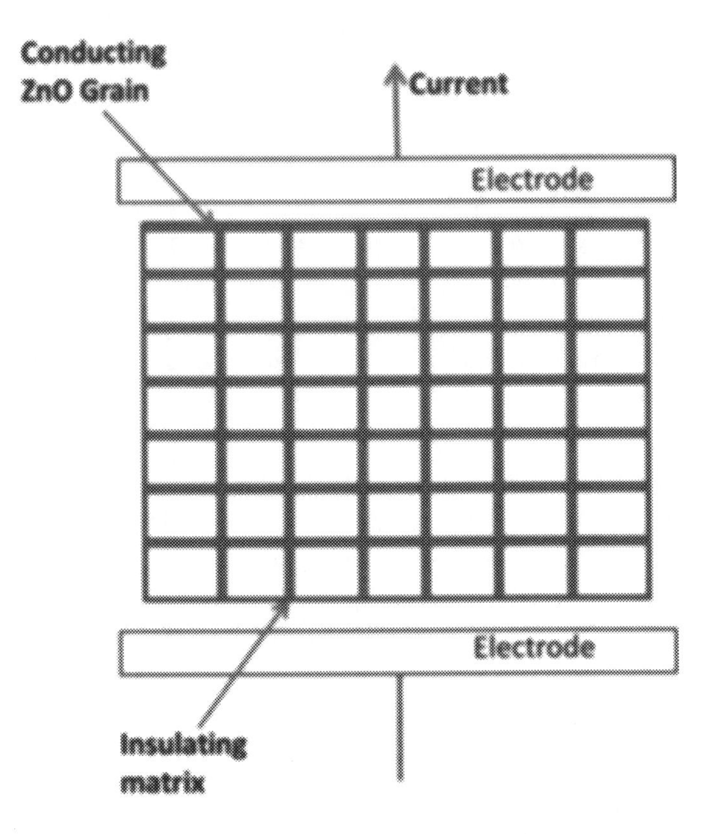

Fig. 6: Function of capacitor exploring conducting ZnO grains in insulating matrix.

Under consideration of distribution of charges in a ferroelectric ceramic crystal (not grain), if it is not symmetric about the crystal's center, a shift in polarization can be produced by deforming the crystal. Thus, mechanically deformed ceramics develops electric charge. This is also an example of influence of atomic structure in the macroscopic property of piezoelectric ceramic. Thus, piezoelectric ceramic powerfully convert mechanical energy into electrical energy or vice versa using piezoelectric ceramic.

High-temperature ceramic oxide superconductors:

Superconducting ceramics find a variety of applications of energy technologies and sensories. Table 4 summarizes selected and important contemporary and potential applications of superconductors

Table 4: Applications of Superconductors

Sensories	Ultra-fast optoelectronics	Energy techniques	Magnetic techniques
SQUID magnetometers MIR sensor coils IR bolometers X-ray detectors	Photodiodes Telecommunication Microelectronics Transistors Digital switching	Power cables Generators, motors, transformers	Accelerator magnets, fusion magnets, Ultrahigh-field magnets NMR, magnetic levitation trains, wind turbines, flying wheel

Where

SQUID=superconducting quantum interference device
MIR = Micropower impulse radar.

With these enormous applications of today superconducting ceramics, the rest of this section covers

a. Definition
b. classification
c. crystal chemistry of cuprate superconductors
d. theory &
e. descriptions of selected applications.

a. Definition:

The ceramic materials that exhibit zero electric resistivity and a strong diamagnetism to below a critical temperature are called superconducting ceramics. The critical temperature ranges from liquid Helium to liquid Nitrogen temperature. The diamagnetism means expulsion of magnetic flux lines.

b. Classification:

One common classification of superconducting ceramics is based on their response to a high magnetic field.

Type 1:

Pure metals except V, Tc and Nb with perfect crystal lattice belong to type 1. Magnetic flux lines are not able to penetrate the material (type 1) and hence, above a critical magnetic field the superconductivity abruptly disappears. Type 1, metal superconductors are not suitable for high magnetic field applications due to lose their superconducting below critical temperature in magnetic field.

Type 2:

Superconducting alloys (ductile) and oxide superconductors such as cuprates (brittle) belong to type 2 superconductors. The type 2 superconductors exhibit superconducting property even under magnetic flux lines. Magnetic flux lines can penetrate the materials while maintaining superconducting property below critical temperature.

Crystal chemistry of cuprate superconductors:

Cuprate superconductors have been playing a vital role in the development of superconductors in general and understading of cuprate superconductors in particular. In fact, discovery of superconducting property of cuprate superconductors won the noble prize in physics in 1987. Table 5 below summarizes some examples of cuprate supercondcutors with their critical temperatures.

Table 5: Examples of superconductors with their critical temperature, K

Superconductor	Critical Temperature, K
$Sr_{0.88}Nd_{0.14}CuO_2$	40
$Sr_{0.9}La_{0.1}CuO_2$	43K
$YBa_2Cu_3O_{6.9}$	93K
$Bi_2Sr_2Ca_1Cu_2O_{8+y}$	110
$Bi_2Sr_2Ca_2Cu_3O_{10+y}$	125
$HgBa_2Ca_2Cu_3O_8$	134

In the cuprate superconductors, copper ions play a vital and remarkable role in achieving zero resistivity below the critical temperature. Thus, copper atom to vary its charge and oxygen coordination is responsible for the exceptional superconducting properties. Fascinatingly, cuprate structures of superconductors are related to cubic $SrTiO_3$ pervoskite structure.

Large cation such as Ba, Sr or La present in cuprate superconductors occupy at Sr site of 12-coordination where as small cation, copper at the corner of Ti^{4+} site with six coordination environment. When smaller cation such as Ca^{2+}, Y^{3+} is present in the superconductors, it occupies at the strontium site by removing the central oxygen layer from the perovskite structure to form smaller coordination of cations.

$La_{1-x}Sr_xCuO_{4-y}$ belongs to K_2NiF_4 sturcture, which can also be derived from composite structure of interchangeably stacked perovskite and rocksalt type structures.

Remarkably, superconducting structure of $YBa_2Cu_3O_7$ (orthorhombic) is different from that of tetragonal $YBa_2Cu_3O_7$ structure.

The main and essential difference between orthorhombic and tetragonal $YBa_2Cu_3O_7$ phaseses arises from coordination around copper ion present in the structures. Thus, non-superconducting tetragonal $YBa_2Cu_3O_7$ has dumbbell coordination whereas superconducting orthorhombic $YBa_2Cu_3O_7$ structure has four coordination by oxygen uptake under formation of chains and thereby formal charge of copper is supposed to be $Cu^{2+/3+}$.

Theory for superconducting cuprates:

According to cooper pairs theory, coulombic repulsion between electrons are overcome and thereby, cooper pairs of electrons binding to gather takes place. Thus, cooper pairs of electrons binding does not obey Pauli's exclusion principle, which occurs merely at low temperature, i.e. below the critical temperature.

Meisner Effect:

When a material turns to be superconductors at or below critical temperature, the superconducting material excludes a magnetic field from it. This effect is called Meisner effect.

Description of applications:

Enormous applications of superconducting ceramics in addition to a thin layer and single crystals are outline below.

Cables for power Grids:

Because of development of fabrication of kilometer-long superconducting cables and ribbons, fivefold more efficient than those operated with conventional copper wires is achieved. The cables can be installed underground and transformation of conventional current can be eliminated.

Superconducting Magents:

Since superconducting magents achieve high current densities in excess of 100 A/mm^2, their applications range from low-field magnets to ultrahigh field NMR magents with frequencies exceeding 1 GHz.

In addition, application of superconducting magnets is extended to MRI instruments used extensively in diagnostic medicine. Equally, generators for automotive applications can benefit from superconducting magnets.

Other applications of superconducting ceramics are magnetic shielding, magnetic energy storage and magnetic levitation and combined bearings etc.

Ferromagnetic semiconductor ceramics:

Effect of atomic structure on macroscopic property of ceramics is illustrated with a novel and new ceramics, ferromagnetic semiconducting ceramics. Thus, Ferromagnetic semiconducting ceramics explore combination of ferromagnetic property and semiconducting property, which makes an unique and novel property of ceramics. In this ceramics, spin current is also explored. It is noted that spin current is different from charge current. The table 6 below summarizes differences between spin current and charge current.

Table 6: Charge Current and Spin Current differences

Charge Current	**Spin Current**
Easy to generate	Generation is difficult
Easy to transport	Transportation is difficult
Easy to detect	Detection is also difficult

However, in the modern and revolutionary world, spin current is able to generate and its transportation and detection is possible.

Characteristics of Spin Current:

1. Ohmic resistivity is suppressed.
2. No electrons, no charges transported.
3. Invarient under time reversal
4. Spin flows.

Half-metallic ferromagnet shows spin current generation. Thus, one spin of it behaves as semiconductor and another spin with opposite behaves as metals. Thus, for ferromagnetic materials, even number of spins with opposite direction is required.

Magneto resistance:

Magneto resistance is defined as follow.

$$\text{Magnetoresistance} = \frac{\text{Resistance without magnetic field}}{\text{Resistance with magnetic field}}$$

ΔR = I Resistance without magnetic field — Resistance with magnetic field I Magnetic resistance is defined as ratio of ΔR to R.

Bioceramics:

Bioceramics are one of the major demanding applications among advanced ceramics in medicine. Thus, bio-inert ceramics such as alumina, zirconia and titania are explored in addition to bioconductive ceramics such as hydroxyapatite tricalcium phosphate and calcium phosphate ceramics. These bioceramic materials find clinical applications in joint replacement, bone grafts tissue engineering and dentistry. Examples of metallic, ceramic and polymeric biomaterials and their applications are summarized below in the table 7.

Table 7: Applications of engineering materials

Biological Behavior	Material	Applications
Ceramics:		
Bio-inert	Alumina	Femoral balls, inerts of acetabular cups,
	Yttria Stabilized Tetragonal Zirconia	Femoral balls
Bio-active	Hydroxyapatite	Bone cavity fillings, coatings, ear implants, vertebrae replacement
	Tri-calcium phosphate	Bone replacement
	Tetra-calcium phosphate	Dental Cement
	Bio Glass	Bone replacement
Metal/Alloy		
Bio-inert	Titanium metal	Acetabular cups
	Ti6Al4V alloy	Shafts for hip implant, Tibia
	CoCrMo alloy	Femoral balls and shafts, Knee implants
Polymer		
Bio-inert	Carbon (Graphite)	Heart Valve Components
	High Density Polyethylene	Articulation components

Bio inert materials do not release any toxic constituents in addition to not showing any positive interactions with living tissue. Bio active materials do show a positive interactions with living tissues.

Bio active hydroxyapatite:

Synthetic hydroxyapatite is very strictly related to naturally occurring biological apatite bones with respect to its structure and chemical in nature. Therefore, extensive and elaborate work has been carried out on synthetic and bio active hydroxyapatite. In this section, structure, crystal chemistry and biomedical function of synthetic hydroxyapatite are the subject of topical interest.

Crystal chemistry:

It is essential and extremely desirable to understand the crystal structure and chemistry to invoke the important and interesting properties of mechanically favorable bio active hydroxyapatite.

General formula with several compositions but structurally identical is $M_{10}(ZO_4)_6X_2$

Where M = Ca, Pb, Cd, Sr, La, Ce, K & Na
Z = P, V, As, Cr, Si, C, Al &
X = OH, Cl, F, CO3, H2O.

However, hydroxyapatite has the formula $Ca_{10}(PO_4)_6(OH)_2$, which can be derived from various possible compositions.

Coordination and arrangement of every ions present in the hydroxyapatite are described first followed by formation of unit cell will be highlighted.

Coordination Environment around cations of hydroxyapatite:

As usual PO_4 tetrahedra are the coordination arrangement around phosphors atom. Calcium atom forms two different coordination environments and thus, Ca_1 atoms have nine coordination environment and Ca_2 atoms have irregular six coordination environment. Phosphate groups, PO_4 contribute their oxygen atoms for the coordination around Ca_1 and Ca_2 atoms.

Three different polyhedral present in the hydroxyapatite lead to unique and porous but mechanically strong structural feature for the special type of bioactive hydroxyapatite. Calcium polyhedral share faces to form chains parallel to crystallographic C-axis. The two parallel chains of Ca1 and Ca2 polyhedral are linked such that hexagonal array with PO4 tetrahedral is formed. The hydroxyl groups are found to occupy the hexagonal channels.

Hydroxyapatite Coatings for Biomedical applications:

Titanium plate is known metal for bone implant. But, to make easier for bone tissue growth hydroxyapatite is regularly coated on titanium plate for practical and medical implants. Thus, bone tissue willingly and certainly grows onto hydroxyapatite coated titanium plate. Therefore, intensive research is being focused on formation of hydroxyapatite thick coatings by plasma spraying.

Performance requirements of bioceramic hydroxyapatite coatings are summarized below in the table 8. These coating properties help decide for the selection of implant applications.

Table 8: Requirements of Bioceramic Hydroxyapatite coatings

Property	Requirement	Reason
Coating thickness	➢ 50 micron	Coating will be resorbed
	< or = 200 micron	Upper limit of sufficient strength
Hydroxyapatite content	➢ 95%	Minimum purity of biocompatibility
	➢ 98%	Chemical stability
crystallinity	➢ 90%	Increasing resorption with decreasing crystallinity
Adhesion strength	➢ 35 MPa	Prevention of spalling
Porosity/Toughness	100 micron	Minimum porosity for ingrowth.

In addition to naturally found hydroxyapatite, man -made materials such as calcium-titanium-zironium phosphates are being studied for their better properties and performance to replace hydroxyapatite coatings. Thus, $CaTiZr_3(PO_4)_6$ is an example for the purpose of detailed investigations. Thus, solubility of $CaTiZr_3(PO_4)_6$ is lower than hydroxyapatite and adhesion property of orthophosphate is better than hydroxyapatite. The good coating of orthophosphate is studied by atmospheric plasma spraying powder.

Magnetic Ceramics:

Ceramics have been played a major role in the development of applications of magnetism. Thus, commercial interest and application of magnetic ceramic started only in the early 1930's using copper and cobalt ferrites. However, the first and ever commercial magnetic ceramics were produced in 1952 by Philips company researchers.

Origin of Magnetic dipole in magnetic materials:

Electric current, due to movement of electrons, gives rise to magnetic force. As all the materials are made up of atoms/ions, orbital motion and spin of electrons in each atom present in the material results in the production of macroscopic magnetic properties of a material. Between orbital motion and spin of electrons, magnetic moment due to spin of electrons is dominant over orbital motion. Therefore, theoretical calculation of magnetic moment of cations for examples, Ti^{3+}, Cr^{3+}, Fe^{3+}, Cu^{2+} involve essentially from spin of electrons present in the ions. Table 9 below summarizes electronic configuration, calculated magnetic moment and measured magnetic moment for a few first row transition metal cations.

Table 9: Comparison of calculated and measurement magnetic moment of first row transition metal cations

Cation	Electronic Configuration	Calculated Magnetic Moment	Measured magnetic moment
Ti^{3+}	$3d^1$ (one unpair electron)	1.73	1.8
Cr^{3+}	$3d^3$ (three unpair electrons)	3.87	3.8
Fe^{3+}	$3d^5$ (five unpair electrons)	5.92	5.9
Co^{3+}	$3d^7$ (two unpair electrons)	3.87	4.8
Cu^{2+}	$3d^9$ (one unpair electron)	1.73	1.9

The five classes of magnetic materials are known and these are

(i) diamagnetic ceramics
(ii) paramagnetic ceramics
(iii) ferromagnetic ceramics
(iv) antiferromagnetic ceramics and
(v) ferromagnetic ceramics.

Two of them are outline very briefly below.

Diamagnetic ceramics:

As described in the bonding of ceramics (typically covalent bonding is present in ceramics), all the electrons are paired by covalent bonding results in diamagnetic of ceramics commonly. The pairing of electrons lead to net magnetic moment due to electron spin is zero. Therefore, most diamagnetic ceramics do not find applications in commercial uses and a least interest is found in scientific as well.

Paramagnetic ceramics:

Paramagnetic ceramics exhibit magnetic moment due to presence of unpaired electron spins. Therefore, magnetic susceptibilities are positive. Most first row transition metals are paramagnetic because they have unpaired electrons in their 3d orbitals. Also, non-transition metals such as Na may be paramagnetic due to the alignment of the spin moment of the valence electrons with the applied field, this effect is known as Pauli paramagnetism.

Classes of Magnetic Ceramics:

Three classes of magnetic ceramics which are based upon ferrites are structurally and compositionally as well different. Table 10 below summarizes the classes of magnetic ceramics.

Table 10: Classes of Magnetic Ceramics

Structure	Composition	Applications
Spinel	$MeFe_2O_4$, Me = Ni, Co, Mn, Zn	Soft magnets
Garnet	$RE_3Fe_5O_{12}$, RE = Y, Gd	Microwave devices
Magnetoplumbite	$MeFe_{12}O_{19}$, Me = Ba, Ca, Sr	Hard Magnet

Hard Ferrites:

Hard ferrites are found applications in

(1) starter and motors in automobiles
(2) Loud speakers
(3) Rotors for cycle dynamos
(4) windscreen wiper motors
(5) Decorative magnets for refrigerators
(6) DC motors in fuel pumps
(7) electric shavers, food mixers and coffee grinders and
(8) magnetic strips on credit cards, ATM cards etc.

These applications of hard magnetic ceramics are based upon their permanent magnets in nature. These hard ferrites, having permanent magnets, have characteristics of large Hc and Bc. To have such a large Hc and Bc, ferrites with magnetoplumbite is the preferred choice.

Magnetic ferrites are preferred over metallic magnets due to the following valid reasons.

(1) Raw materials are comparatively cheap and widely available
(2) manufacturing processes are simple.

Soft Ferrites:

Soft ferrites find plenty of household applications and some of them are

(1) magnetic recording and data storage media
(2) Transformer cores in telephones
(3) line transformers, deflection coils, tuners and rod antennas in radios and televisions.

Soft ferrites have high magnetic moment, low Hc and narrow hysteresis loops. Spinel ferrites with cubic structures exhibit the properties of soft ferrites.

Microwave ferrites:

Microwave ferrites find applications such as

(1) radar absorbing paint
(2) radar phase shifters.

Other applications of magnetic ceramics:

(1) Audio and video cassette tapes
(2) floppy disks
(3) computer hard disks
(4) credit cards.

Optical ceramics:

Optical ceramics include

(1) ceramic pigments and stains
(2) translucent ceramics
(3) lamp envelopes
(4) fluorescence
(5) optical fibers
(6) phosphors and emitters
(7) solid state lasers.

Questions:

1. What are ceramics?
2. How do advanced ceramics differ from traditional ceramics?
3. What are types of functional ceramics? Provide at least two examples on each type.
4. What is Garnet and give an example for garnet type ceramics
5. What are the applications of Silicon carbide?
6. How does crystalline SiO_2 differ from glass SiO_2?
7. What are point defects?
8. What are Schottky defects?
9. What are Frenkel defects?

10. What is brittle fracture?
11. How does tensile strength of alumina differ from compression strength of alumina?
12. How does electrical conductivity vary with temperature for VO_2?
13. What is varistor?
14. What is ceramic superconductor?
15. What is the maximum critical temperature of cuprate superconductor achieved?
16. What is Meisner effect?
17. What are the applications of HTC superconductors?
18. What is ferromagnetic semiconductor ceramics?
19. What is magneto resistance?
20. How does bio inert ceramic differ from bioactive ceramics?
21. What is the structure of hydroxyl apatite?
22. What are the essential properties to be satisfied for bioceramics?
23. What is the calculated magnetic moment of Co3+?
24. Define and describe types of magnetic ceramics.
25. What are hard and soft ferrites?

Cement, Refractories and Abrasives

Objectives:

1. To emphasize the need for this chapter.
2. To outline history of cement
3. To define a general process involving cement product
4. To elaborate chemistry of cement formation
5. To outline kinetics of formation of phases during the cement production.
6. To describe chemistry of cement hydration.
7. To outline the mechanism of calcium silicate hydrate (CSH) gel formation by NMR studies.
8. To elaborate chemical corrosion of cement
9. To elaborate Portland cement phases.
10. To introduce refractories with definition and prime requirements of refractories.
11. To classify the refractories based upon operating temperature
12. To define classification of refractories based upon chemical composition
13. To elaborate properties of refractories.
14. To describe manufacture of refractories.
15. To define and classify abrasives

16. To elaborate physical properties of abrasives
17. To define synthetic refractories
18. To define SiC refractory
19. To define BN abrasive.

Cement is a more common material but it has vital importance in many areas of lives in the earth. However, cement has a rich in chemistry and hence, in this chapter, it is aimed to address the following questions. The common but vital questions are

1. Why is optimal composition required to achieve the best properties of any kind of cement?
2. What is cement setting? And how does it work?
3. What are the types of processes required for cement setting?
4. How long does the cement setting take place?

Introduction:

Cement was the first structural material exploited using its three basic components such as sand, lime and water. it essentially consists of polymerized and dense calcium silicate hydrates paste at the molecular level. Its main and unique property is ability to to set and remain insoluble material in water. Because of its property of setting associated with cement, it finds application to bind large stones and bricks. Concrete also can be made using cement by mixing with sand and stones.

History of cement:

Ancient Romans started to produce mortars by mixing of lime, volcanic ash and crushed clay. This type of cement was known as Pozzolanic cements. Later, it was confirmed that the Pozzolanic cements attain strength from aluminate phases present in the volcanic ash, which promotes hydration of the final cement powders.

In Europe, due to the limited availability of volcanic ash art of production of cement from the precursors was inconsistent. Therefore, until 1756, in fact, the development of chemistry of cement did not make any breakthrough. When Eddystone lighthouse in Cornwall, England was considered to rebuild, Smeaton discovered better and superior results from the experiments with impure limestone. The superior results lead to strong cement, which would be equal to Portland cement properties.

The advancement of cement occurred due to the expansion of uses of cement especially, large construction projects in addition to availability of characterization techniques to understand chemistry of cement, which ultimately helped to improve the properties of cement.

In general, two major steps/processes are involved in the cement preparation. The first step is the high temperature mixing and processing of raw powders such as limestone, sand and clay to get cement powder. The second step is the hydration, mixing and setting of cement powder into final cement product. For example, the dry portion of Portland cement consists of about 67% calcium oxide, 20% silica, 6% alumina, 3% iron(III) oxide and small amount of other materials. The two processes involving cement product preparation are shown in the equations below.

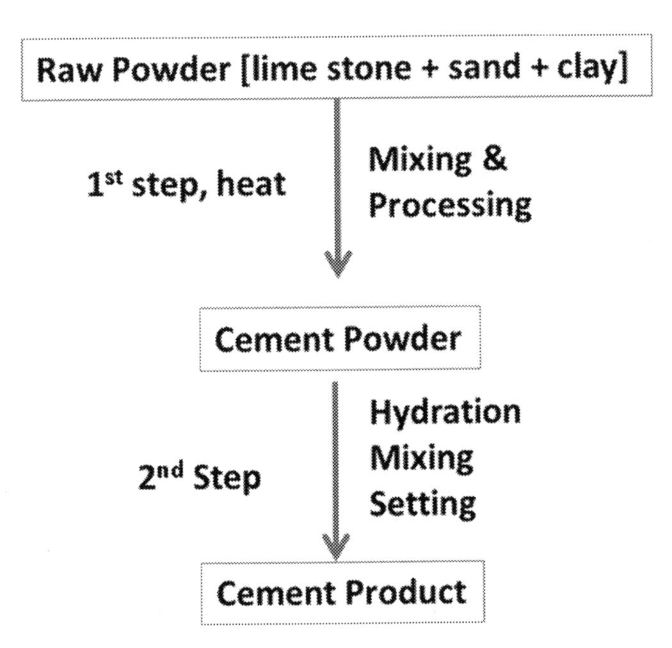

Chemistry of cement Formation:

a. Preparation of cement precursors-Clinkers:

The first and the foremost step is the preparation of cement precursors. The preparation of cement precursors involves grinding the required amount of raw materials and heating the ground powders to high temperature while rotating the mix. The mixing helps the raw powder to be homogeneous and heating to high temperature results in evaporation of water and release of carbon dioxide, a decomposed product. Also, at high temperature, melting is observed with reaction between raw powders. Finally, cooling ends with formation of clinkers. Clinkers mean stony, heavily burnt material. Thus, clinkers are essential for cement production. However, clinkers are again subjected to grinding for the production of cement. With these overall steps involved in the manufacture of cement, more detail study on chemical constituents and their function including phase formation will be described blow. Three main chemical constituents are found importance for the cement production. These are

1. Calcium oxide
2. Silica and
3. alumina.

When these three major chemical constituents are subjected to cement product formation by two major steps described earlier, presence of two major chemical components are observed. In fact, the two chemical components certainly play important role in achieving high quality cement powders. The two chemical components are tricalcium silicate, C_3S and dicalcium silicate, C_2S. These are the two chemical compounds responsible for good quality of clinkers formation. The uniqueness of these two calcium silicates is that tricalcium silicate and dicalcium silicate have vigorous chemical reaction property with water to produce cement paste. In addition, between the two types of calcium silicates, tricalcium silicate has higher hydrates reaction rate and has faster setting than that of dicalcium silicate clinkers and hence, tricalcium silicate is a preferable clinker material. However, solid state synthesis of formation of tricalcium silicate requires temperature of 1200 – 1400°C for several days, which turns to be not economical.

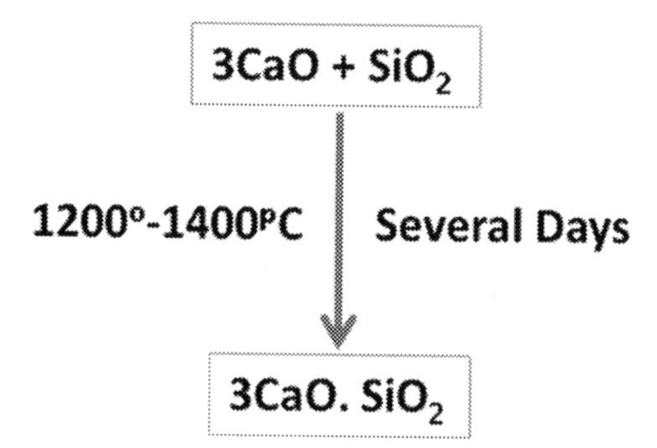

Therefore, synthesis of $3CaO.SiO_2$ from melt-technique is a viable in terms of duration. But, still it requires at least 2200°C, which is also an impractical due to involvement of high temperature.

In order to avoid the high temperature of 2200°C, but retaining the liquidous synthesis of $3CaO.SiO_2$ production, flux type synthesis of 3CaO. SiO_2 is preferred due to economical and practical reasons. Thus, fluxing agent, Al_2O_3 happens to be reducing the synthesis temperature and duration as well. The synthesis temperature is reduced from 2200°C to 1350° – 1500°C only.

The total process of cement clinker focuses on final products of C_3S in large portion, C_2S in small portion. The raw materials such as lime ($CaCO_3$), Quartz (SiO_2), Clay (Al_2O_3) and water are subjected to temperature of maximum 1500°C. Initially, water boils off, and above 700°C, calcium carbonate (lime) decomposes into CaO with releasing CO_2. Then, two moles of total CaO reacts with quartz (SiO_2) to form dicalcium silicate, $2CaO.SiO_2$ and reaction of CaO with clay, Al_2O_3 results in formation of Ettringite phase. The Ettringite phase melts at 1450°C to form liquid phase. The liquid phase, then, favors the rapid production of tricalcium silicate, $3CaO.SiO_2$ at 1500°C with smaller portion of dicalcium silicate, aluminate and alumino ferrate phases remaining in the cement product. The final cement proportion is not get affected by the minor impure iron oxide in the raw materials.

Chemistry of Cement hydration:

Cement hydration is a familiar process but its chemistry is complicated to understand. Cement powder is mixed with water so that 30 – 40% of total mass is water after hydration. This range does not vary significantly with different compositions of cement powder.

The rate of hydration of tricalcium silicate can be divided into four different regions. The four regions are

1. Pre-induction period
2. Induction period
3. Post-induction period and
4. Saturation period.

The properties and chemistry of four regions are described below.

Chemistry of pre-induction period of hydration:

When cement is in contact with water, immediate and intense hydration of tricalcium silicate takes place. The hydration of tricalcium silicate leads to dissolution of C_3S by formation of $Ca(OH)_2$ as shown in a chemical reaction below.

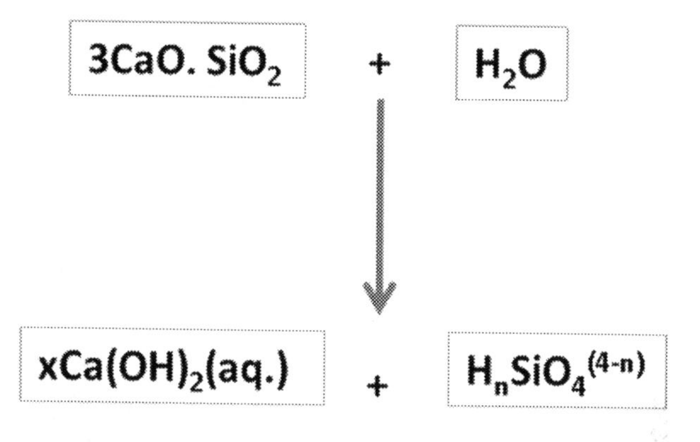

Silicate material of $3CaO.SiO_2$ enters into liquid phase as $H_nSiO_4^{(4-n)}$. The dissolved components Ca(OH)2 and $H_nSiO_4^{(4-n)}$ react to form Calcium silicate Hydrates (CSH) as shown below.

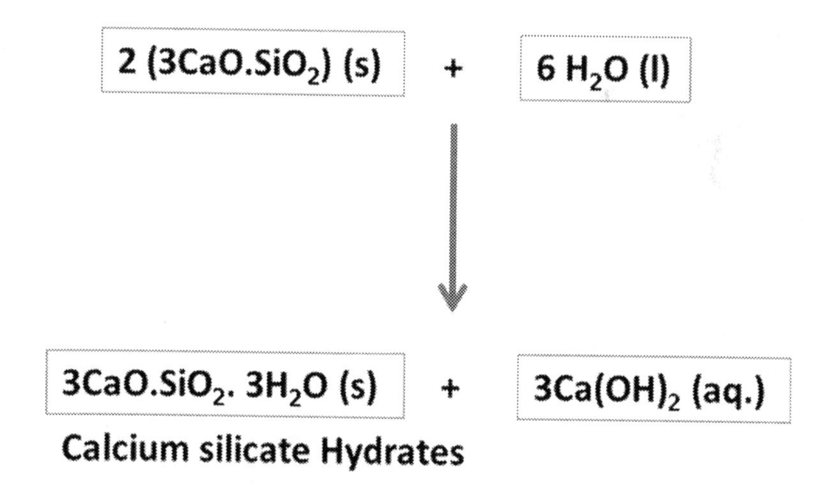

However, the rate of hydration of cement is decreased by the addition of gypsum ($CaSO_4$) prior to packaging so that rapid setting of the cement is mitigated. The gypsum reacts with tricalcium aluminate in presence of water to form various aluminate and sulfoaluminate phases (Ettringite phases) as shown below in the equations.

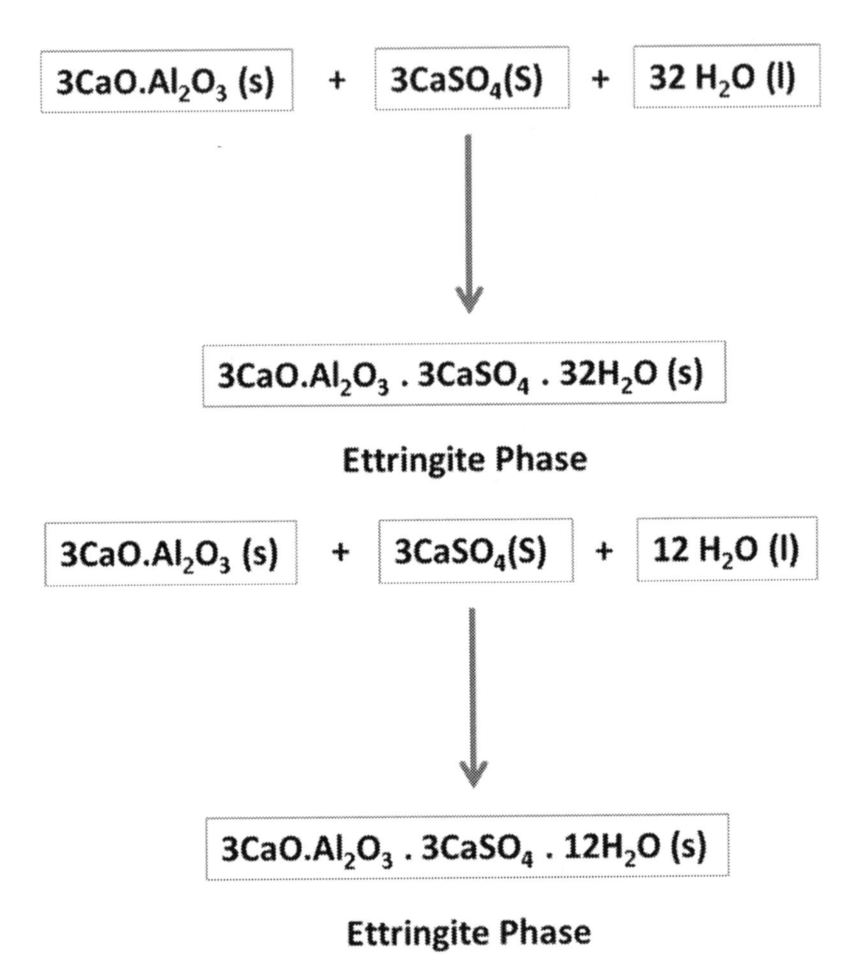

The formation of Ettringite phase continues until they get saturated. Total time of hydration of pre-induction period is a maximum of few minutes.

Chemistry of induction period of hydration of cement:

After a few minutes of pre-induction period of hydration of cement, hydration in the induction period begins with significantly slow.

Several theories have been proposed for the hydration reaction pre-induction and induction periods. One of them stated that rapid hydration in the pre-induction period covers the entire surface of dissolving tricalcium silicate and further rapid hydration is slow in the induction period. Also, another possibility is that formation of $Ca(OH)_2$ product during hydration prevents the rapid hydration of tricalcium silicate in the induction period, which is evidenced by the slope of the curve (II in the graph above).

Chemistry of post-induction period of hydration of cement:

After immediate slow induction period of hydration, accelerated hydration reaction rate is observed. During this period, the hydration products are limited by the nucleation and growth of the hydration reaction. The third period of hydration is the rapid hydration reaction rate. The hydration reaction is represented by the following chemical equation.

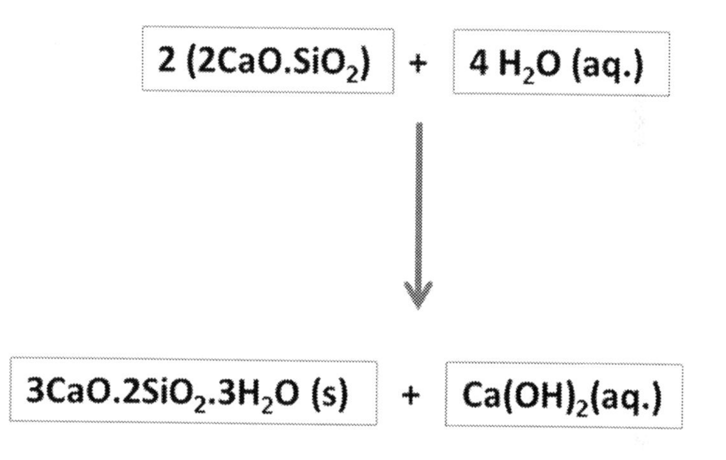

Chemistry of saturation period of hydration of cement:

During this period of hydration reaction, formation and crystallization of $Ca(OH)_2$ reaches at maximum. The crystalline $Ca(OH)_2$ is referred as Portlandite by cement chemists.

NMR studies of Calcium silicate Hydrate (CSH) gel formation:

Formation of calcium silicate hydrate gel is the vital step to understand the process of cement hydration reaction. Thus, solid state nuclear magnetic resonance (NMR) technique is explored due to the consisting of several NMR active isotopes such as ^{1}H, ^{29}Si, ^{27}Al and ^{23}Na.

Among them, ^{29}Si magic angle spinning (MAS) NMR is used to determine silicon – oxygen bonding during hydration reaction of cement, which helps guide the understanding of calcium silicate hydrate gel formation.

Various possible Si – O bondings of silicate unit of cement with ^{29}Si NMR value are shown in the figure 1 below.

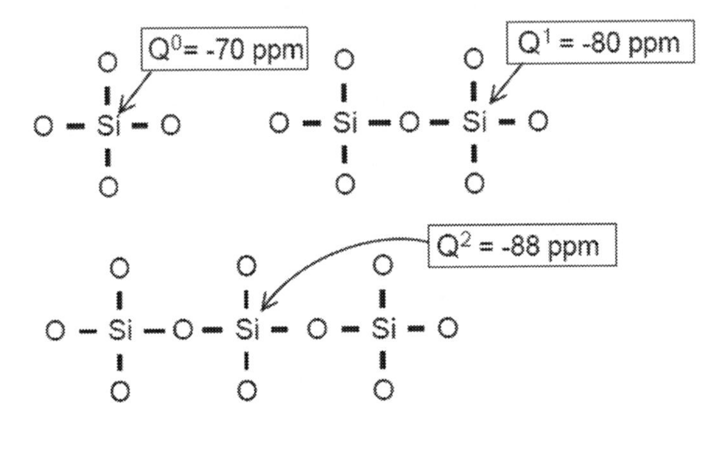

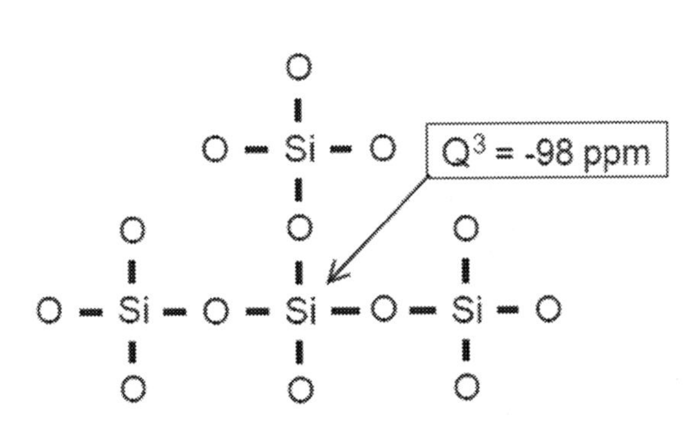

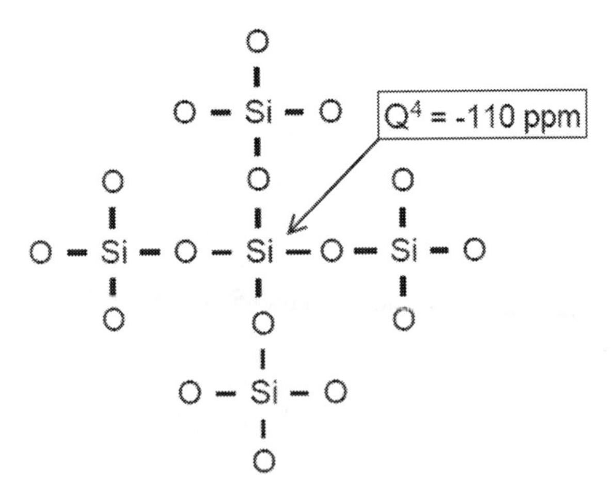

Where Q^0 = monomer unit

Q^1 = Dimer

$Q^2/Q^3/Q^4$ = Polymeric unit.

Fig. 1: Various possible Silicon – Oxygen linkage formation during Calcium Silicate Hydrate gel formation.

During the pre-induction period, presence of monomer unit is observed. With time of hydration reaction, dimer and polymeric units start to appear in the ^{29}Si MAS NMR.

Chemical Corrosion of Cement:

Environmental pollutions are responsible for chemical corrosion cement. For example, acid rain dissolves $Ca(OH)_2$ and hence, it destroys the cement. Another example is the action of CO_2 in the atmosphere on the cement. This leads to conversion of $Ca(OH)_2$ into $CaCO_3$ as shown by step by step chemical reactions.

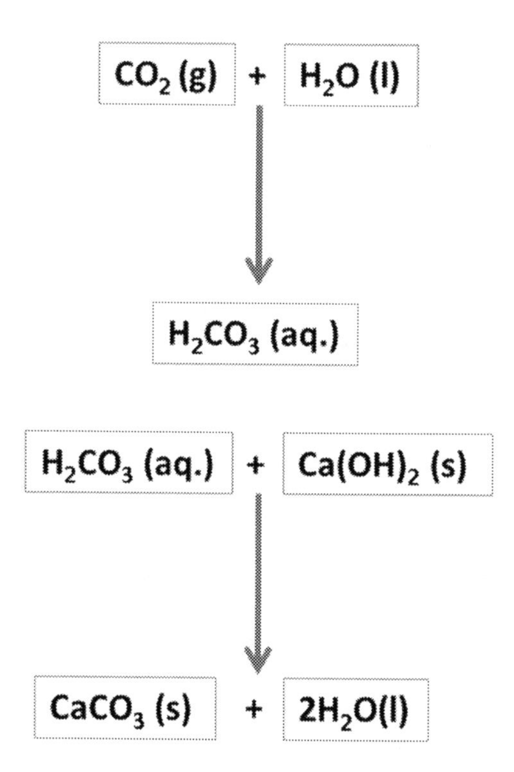

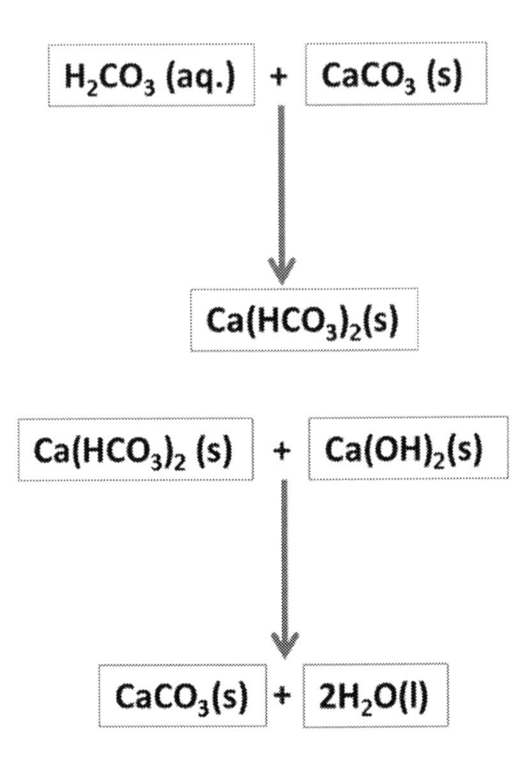

This process leads to deplete the cement of $Ca(OH)_2$ and leaves $CaCO_3$ deposits inside the cement. This process is called carbonation.

Marine environmental pollutions such as $(NH_4)_2SO_4$ (aq.) and $MgSO_4$ (aq.) cause cement degradation by the action of them on the cement as shown below in chemical equations respectively.

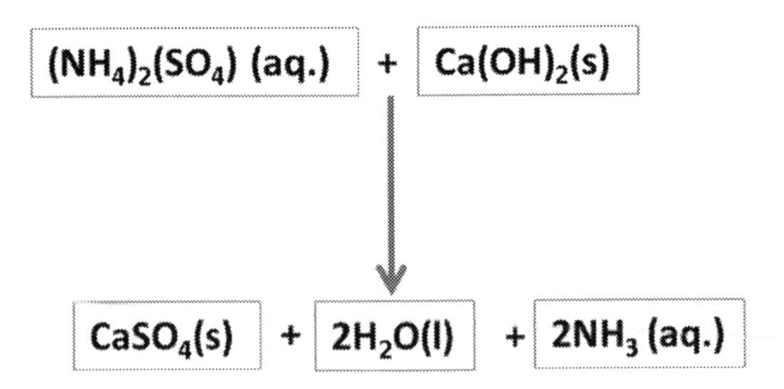

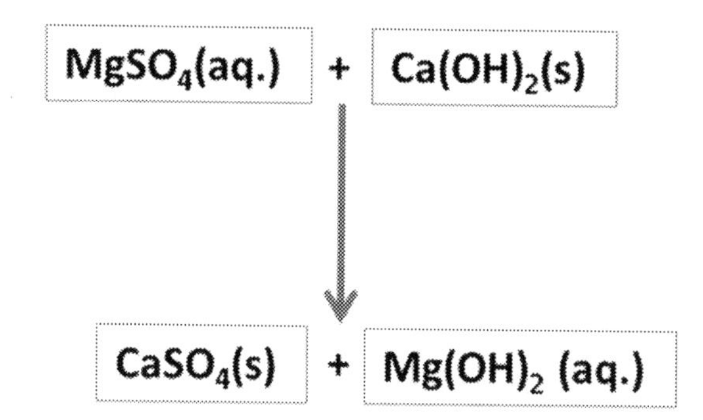

Portland Cement Phases:

Portland cement contains five main components. These are

1. Tricalcium silicate, C_3S, Ca_3SiO_5
2. Dicalcium silicate, C_2S, Ca_2SiO_4
3. Tricalcium aluminate, C_3A, $Ca_3Al_2O_6$
4. Tetracalcium alumino ferrite, C_4AF, $Ca_4Al_2Fe_2O_{10}$ and
5. Sulfate phase (gypsum), $CaSO_4.2H_2O$.

Tricalcium silicate, C_3S (Ca_3SiO_5):

The formation temperature of tricalcium silicate ranges from 1250°C to 2150°C. It is formed by the combination of dicalcium silicate and CaO at above 1250°C. At room temperature C_3S has a triclinic structure. With raise in temperature triclinic transforms into monoclinic and then, into Rhombhohedral phase. Due to metastable nature of C_3S at room temperature, C_3S does not disintegrate into C_2S and CaO below 1250°C. Its dissociation reaction is very slow and in fact, rapid quenching can prevent dissociation reaction and hence, C_3S phase is preserved in triclinic form at room temperature. In order to stabilize high temperature phase of monoclinic at room temperature, foreign ions such as Al^{3+}, Mg^{2+}, Fe^{3+}, Mn^{3+} and P^{5+} have been incorporated in C_3S. These ions are called stabilizers and the stabilizing ions can influence the formation of C_3S phase and its hydration reactivity as well. C_3S can be produced by either by solid state sintering process or by microwave sintering and sol-gel process as represented by a chemical equation.

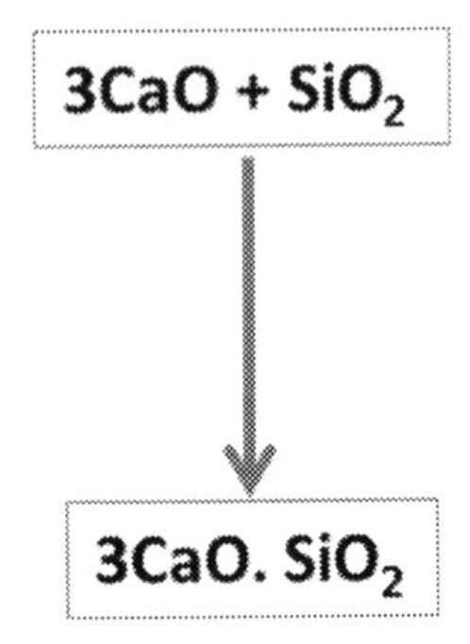

Dicalcium Silicate, C_2S (Ca_2SiO_4):

Dicalcium silicate is a stable form from room temperature to its melting point of 2130°C. However, it undergoes several polymorph phase transitions. Among them, beta C_2S is used in industrial clinkers. Most unwanted industrial clinker is gamma C_2S and this is due to its non-hydration reaction with water. Therefore, it is essential to exclude some unwanted phases in the production of Portland cement in particular and cement in general.

In addition to solid state synthesis of C_2S, sol-gel method is also explored to synthesize C_2S. Thus, it requires calcium nitrate and silicon oxide as starting materials. However, the solid state synthesis of dicalcium silicate is represented by a chemical equation below.

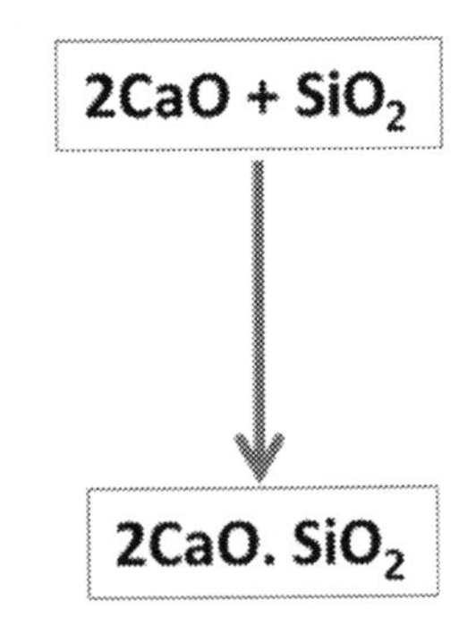

Tricalcium aluminate, C_3A ($Ca_3Al_2O_6$):

Industrially, tricalcium aluminate, C_3A including 2% SiO_2 and 3 – 4% Fe_2O_3 in addition to the most critical substitutent, Na^+ ion.

C_3A is a cubic structure, which does not need any stabilizer. Cubic C_3A is a common clinker in the cement production. However, presence of Na+ might stabilize orthorhombic phase. The solid state synthesis of tricalcium aluminate is represented by a chemical equation below.

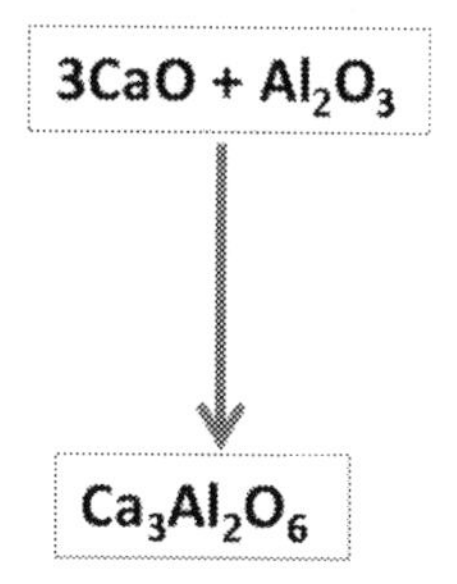

Refractories:

As stated in the preparation of various phases of cement, the temperature of production of cement is required in the order of 1500°C. How can you get into such a high temperature without dissipation/loss for several days in addition to stable reaction container that is useful to manufacture of cement in particular and to high temperature materials in general? The material used in the kiln or furnace is called as a refractory. Therefore, a refractory material is defined that it does not melt at high temperature of operation.

With this brief introduction of refractories, the prime requirements of a refractory are

1. it should not melt at the temperature of a reaction.
2. it should be chemically inert so that it does not involve in the chemical reaction.
3. it should not undergo any structural changes during heating or cooling such that cracking, pitting and splitting can be prevented.
4. Most importantly, it should withstand the thermal shock, sudden and continuous operation of high temperature reaction.

Classification of Refractories:

There are three major classification of refractories based upon the temperature of operation, chemical composition and oxide content.

(i) Classification of refractories based upon the temperature of operation:

The depending upon the temperature of operation of kiln/furnace, the refractory used in the kiln is classified into three types.

(a) Low temperature Refractory:

If the temperature of operation of kiln is less than 1800°C, then, the material used in the kiln as an inner layer is called a low temperature refractory. Thus, the low temperature refractory used in the kiln is fire clay.

(b) Medium temperature Refractory:

If the temperature of operation of kiln is less than 2000°C, the material used in the kiln as an inner layer is called a medium temperature refractory. Thus, medium temperature refractory used in the kiln is chromite.

(c) high temperature Refractory:

If the temperature of operation of kiln is greater than 2000°C, the material used in the kiln as an inner layer is called a high temperature refractory. Thus, high temperature refractory used in the kiln is zircon.

(ii) Classification of refractories based upon their chemical composition:

The depending upon the chemical composition of refractory, the refractory used in the kiln is classified into three types.

(a) Acidic Refractory:

If the major component of the refractory is acidic in nature, then, the refractory material that is useful in the kiln is called acidic refractory. The acid refractory is attacked by alkalis. This type of refractory is stable in acid. Typical examples for acid refractories are silica and zircon.

(b) Basic Refractory:

If the major component of the refractory is basic in nature, then, the refractory material that is useful in the kiln is called basic refractory. The basic refractory is attacked by acid and this type of refractory is stable in basic environment. Typical examples for basic refractories are magnesia and Dolomite (CaO.MgO).

(c) Neutral Refractory:

If the major component of the refractory is neutral characteristics, then, the refractory material that is useful in the kiln is called neutral refractory. Natural refractories belong to neutral refractories. The common examples for neutral refractories are chromite, alumina and zirconia.

(iii) Classification of refractories based upon oxide content:

The depending upon the oxide content of refractory, the refractory used in the kiln is classified into three types.

(a) Simple Oxide Refractory:

If the oxide content of the refractory is simple oxide, then, the refractory material used in the kiln is called simple oxide refractory. Thus, simple oxide refractory used in the kiln is alumina, Al_2O_3

(b) Mixed Oxide Refractory:

If the oxide content of the refractory is mixed oxide, then, the refractory material used in the kiln is called mixed oxide refractory. Thus, mixed oxide refractory used in the kiln is mullite (aluminosilicate).

(c) Non-oxide Refractory:

If the refractory is non-oxide material, then, the refractory material used in the kiln is called non-oxide refractory. Thus, non-oxide refractory used in the kiln is silicon carbide, SiC.

Below is the outline 1 for the three types of classification of refractories.

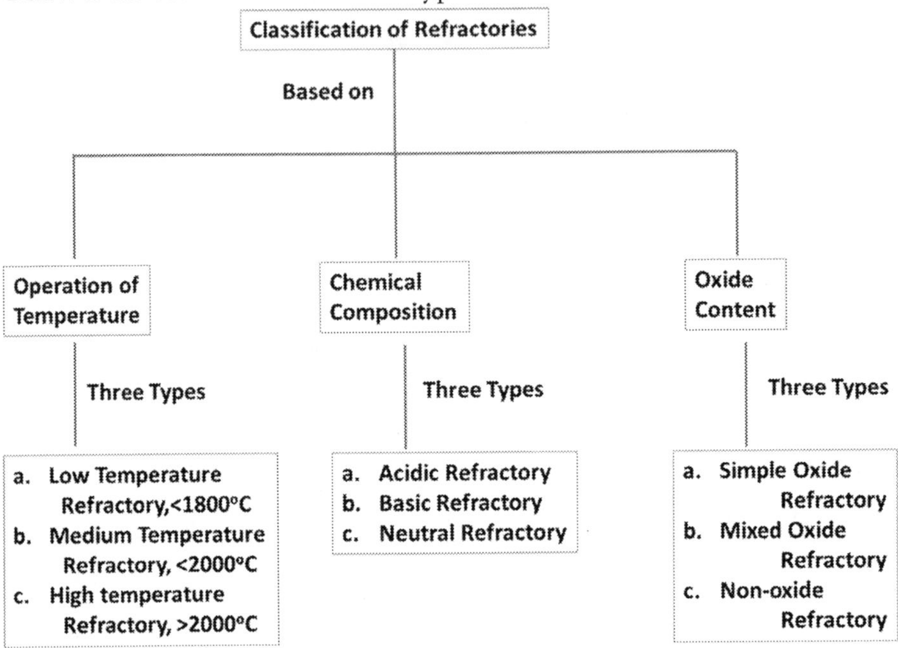

Outline 1: Classification of Refractories

Properties of Refractories:

There is no a single chemical or physical property known to define property of refractories. Therefore, it is always combined several processing to define combo properties of refractories. The properties of refractories are

a. Mechanical properties
b. Thermal properties
c. Thermal conductivity
d. Thermal spallation and
e. Refractoriness.

These properties of refractoriness are outlined below one by one.

a. Mechanical properties of refractories:

The application of refractories decides the mechanical properties of refractories. For example, thermal insulation property of refractory requires highly porous product. For slagging or abrasive conditions, refractory products should be densified one.

Transverse Strength:

Transverse strength is used to find out modulus of strength. It is related to the degree of bonding at room temperature. Thus, fine grained refractories are stronger than coarse-grained refractories. This specifies that low porous refractories are stronger than porous refractories. However, stronger refractories at room temperature may not be having the same strength at the higher temperature due to increase in bond strength by the formation of glassy phase. Also, it is noted that high temperature strength of a refractory is lower than that of room temperature strength of the refractory.

b. Thermal properties of refractories:

Thermal expansion of refractories is an important property to be considered here. Like other solids, thermal expansion is observed for refractories on heating. However, thermal expansion of refractories is much lower than that of metals. It is observed that Thermal expansion of fused silica is very low between room temperature and 1250°C. Whereas thermal expansion of silica increases rapidly from 100°C to 500°C. Then, it becomes stable up to 1250°C. Thus, percentage of linear thermal expansion of fused silica is well below 0.2 % (in the temperature range of 100°C to 1250°C). Percentage of linear thermal expansion of silica is in the order of 1.2% between 500 to 1250°C. Linear thermal expansions of other refractories like MgO, $MgO\text{-}Cr_2O_3$, Cr_2O_3, Al_2O_3, fireclay and SiC fall between the values of silica and fused silica.

c. Thermal conductivity of refractories:

Thermal conductivity of refractories depends upon the chemical and composition of the material. Thus, thermal conductivity of silica is low enough to be a good insulator. Whereas, thermal conductivity if SiC is high enough to be a good thermal conductor. In general, thermal conductivity increases with decreasing porosity.

d. Thermal spallation of refractories:

Refractories are brittle in nature. Therefore, sudden change in temperature either by heating or by cooling can cause cracking and destruction. The raw material and macrostructure of a refractory decide the susceptibility to thermal cracking and spallation of the refractory. However, resistance to thermal cracking and spallation of refractory can be increased by preventing crack formation in the growing of the refractory. In over-all, refractory properties such as low thermal expansion, high density, high thermal conductivity and high strength can exhibit high resistance to thermal cracking and thermal spallation. Typical examples for high resistance to thermal cracking and thermal spallation are fireclay and alumina. Whereas 100% MgO has a poor thermal-shock resistance. However, this thermal shock resistance property of MgO can be improved by the addition of Chrome ore.

e. Refractoriness:

In general refractories do not have sharp melting point due to the mixture of oxides present in the refractories. However, they tend to soften in a temperature range. Thus, refractoriness is defined as resistance to physical deformation under the influence of temperature. Pyrometric Cone Equivalent (PCE) test is usually explored to determine the refractoriness (physical deformation) of refractory at high temperature. Cones are made from finely ground material and it is subjected to heating at a specific rate. The time and temperature needed to bend over and touch the base is compared with that of standard ones.

Manufacture of Refractories:

In general, manufacture of refractories involves several steps from raw materials as shown below in the flow chart 1.

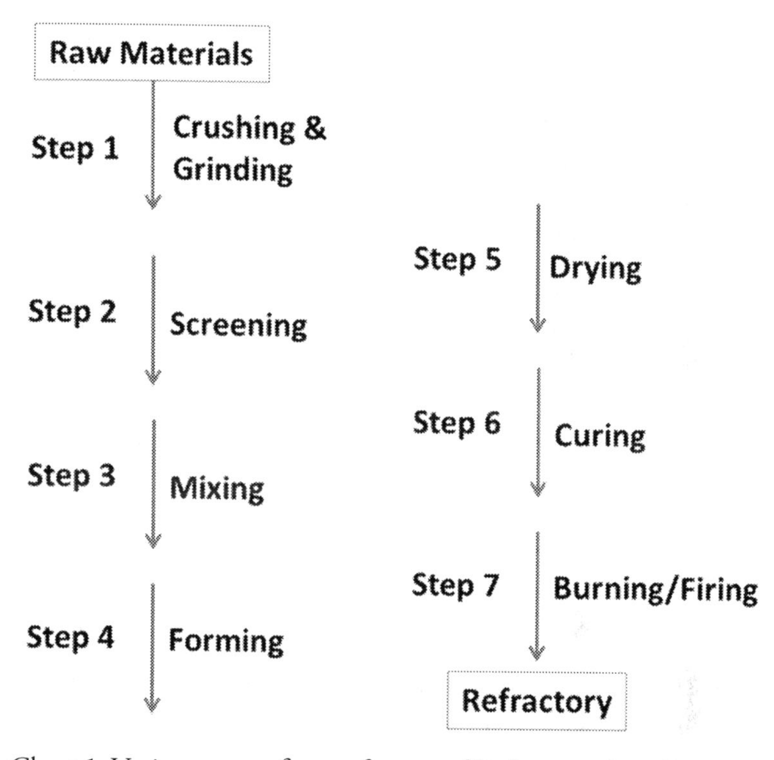

Flow Chart 1: Various steps of manufacture of Refractory from Raw Materials.

Step 1: Crushing and Grinding:

Raw materials such as hard clay and quartzite are needed to crushing process before subjecting the raw materials into grinding for fine grain powder. However, the raw materials of all the refractory are needed grinding step unlike crushing of particular hard raw materials.

Step 2: Screening:

Screening steps is essential in order to obtain sized powder so that high density refractory can be achieved in the final sintering process. Thus, ground powder from step 1 is screened through sieves. Powder that is passed through the sieve will be used directly in the step 3. The remaining powder is returned to grinding process (step 1) until the powder passes through the sieve for the step 3. It is also exceedingly possible to grade the size of the powder screening through different pore sizes of sieves. The vital goal in the screening step is narrow down the particle size distribution that is required for homogenous mixing in the step 3 and for burning step 7 (or sintering process).

Step 3: Mixing:

When more than one material involves in a refractory it is important to mix them homogeneously to get uniform refractory after burning in the step 7. Mixing is also referred to blending. Thus, mixing involves either dry powders or 3 – 8% binding liquid.

Step 4: Forming:

Different shapes of refractories are formed by mechanical equipment. In this operation, a mold cavity is filled with the mix. Sometime, vacuum is added in this step to de-airing the forming so that it promotes a denser product. Also, pressing technique is introduced in this forming step.

Step 5: Drying:

The drying step is critical principally in the large shapes. Sometime, a temperature controlled floor heated by steam is needed in the drying step. Microwave and infrared drying are being investigated to explore the drying process for the better acceleration step.

Step 6: Curing:

Curing temperature is generally higher than the drying temperature used in the earlier step of 5 but it is well-below the burning process of the next step. In this step, strong bonding can be developed by the heat-treatment temperature above 600°C. Thus, adequate strength can be formed to be handled in the burning step.

Step 7: Burning or Firing:

In the burning step in the furnace at higher temperature a ceramic bonding within the refractory is developed and it leads to attain certain desired properties. The firing or burring temperature ranges from 1000 to 1700°C. When gas, oil or coal is being explored as a fuel to get such high temperature, electrical kiln or furnace can also be used to get such a high temperature.

Abrasives

The material that finds applications in grinding, wear down, rub away, smoothing, cleaning or polishing is called abrasive. Therefore, abrasive material is regularly characterized by its hardness. Often abrasive is combined with a binder to make grinding wheels.

Classification of abrasives:

Abrasives are classified into two types based upon their availability. The two types are natural abrasives and synthetic abrasives. Table 1 below summarizes some examples that are known for natural abrasives and synthetic abrasives.

Table 1: Examples for natural abrasives and synthetic abrasives

Natural Abrasives	Synthetic Abrasives
Quartz, Garnet oxide, Corundum and emery and diamond	Fused aluminum oxides, fused ZrO_2 – Aluminas, sintered aluminas, silicon carbide, boron carbide and cubic boron nitride

Classification of abrasives is outlined below.

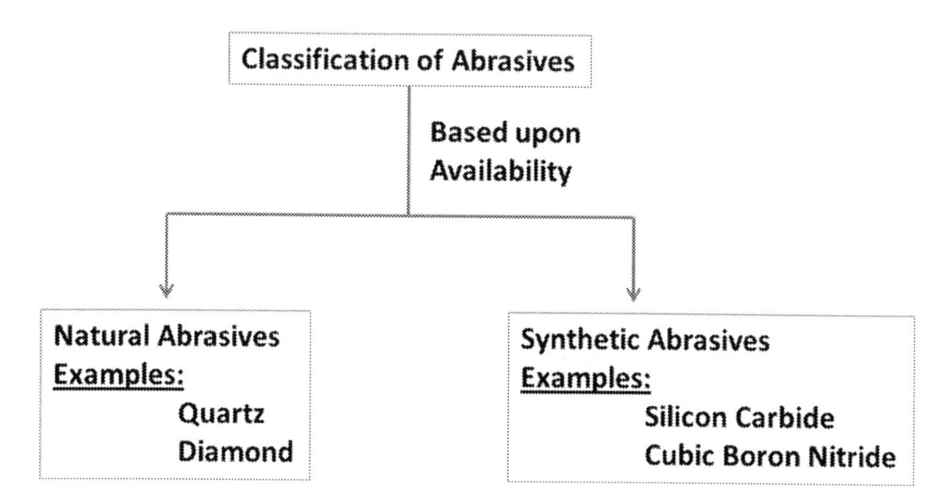

Outline 2: Classification of Abrasives

Another classification of abrasives is based upon their hardness. Thus, the two different abrasives in this category are hard abrasives and soft abrasives (outline 3). Table 2 below summarizes some examples that are known for hard abrasive and soft abrasive.

Table 2: Few examples of hard abrasive and soft abrasive

Hard Abrasive	Soft Abrasive
Diamond Silicon Carbide Aluminum Oxide Garnet Emery (Mixture of $Al_2O_3 - Fe_2O_3$)	Some metal oxides (Fe_2O_3. nH_2O, Fe_3O_4, Cr_2O_3

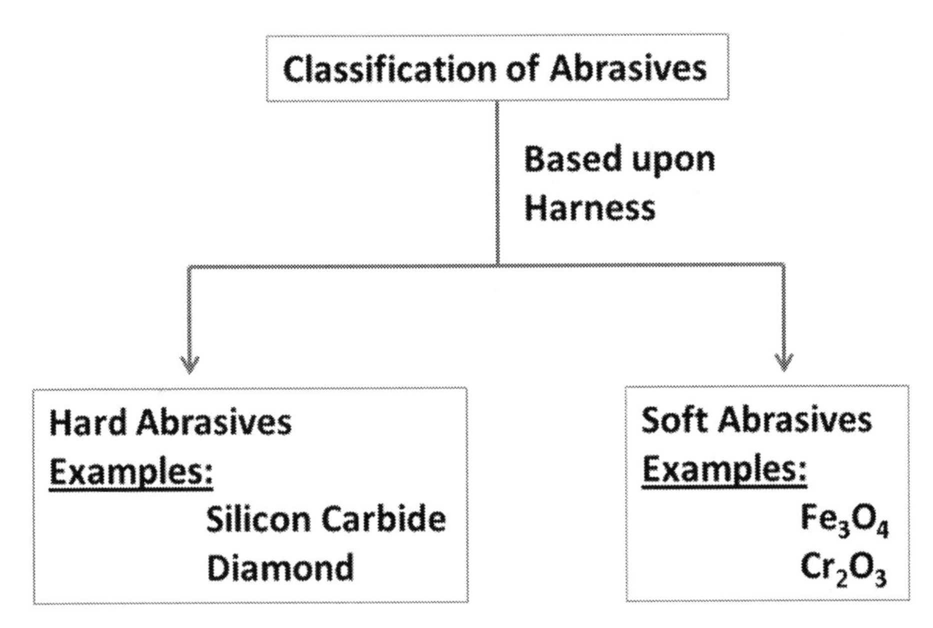

Outline 3: Classification of abrasives based on hardness

Physical Properties of Abrasives:

There are two important physical properties of abrasives considered and these are

a. Hardness
b. Fracture Toughness or Grain Strength

a. Hardness:

Commercial expectation of abrasives should have hardness at least to that of quartz. The Knoop indenter is used to determine hardness. Thus, hardness is obtained by the capacity of a substance to make marks on other substances. In abrasives, Knoop hardness is considered. Whereas Mineralogists consider Mohrs' scale for the hardness of various abrasives. Table 3 below compares hardness of various abrasives.

Table 3: Comparison of various abrasives hardness

Abrasive	Knoop Hardness	Mohrs' Hardness
Quartz (SiO_2)	820	7
$MgAl_2O_4$	1270	-
Garnet	1360	8
Fused Al_2O_3 (White)	2050	9
Silicon Carbide	3050	-
Boron Carbide	3800	-
Cubic Boron Nitride	4700	-
Diamond	7000 – 8000	10

b. Fracture Toughness or Grain Strength:

Fracture toughness is refereed to resistance to crack propagation. Grain strength depends upon

1. Grain shape
2. Micro cracks present in the grains
3. Intrinsic strength of the body.

Fracture toughness is typically determined by studying the crack formation by Vickers indenter. Thus, fracture toughness is commonly calculated by measuring the average crack length and diagonals of the indentations.

Natural Abrasives:

Quartz
Garnet
Corundum and Emery
Diamond

Quartz Abrasive:

It is found worldwide. It has application as abrasive for the finishing of non-metallic substances. The non-metallic substances include leather and felt. The quartz (SiO_2) has a major drawback to be used continuously due to inhalation of silica dust that is formed during its use. Therefore, quartz abrasive can be without difficulty replaced by synthetic abrasives such as fused aluminum oxide and silicon carbide.

Garnet Abrasive:

Garnet abrasive has general formula of $A_3B_2C_3O_{12}$. A typical example of natural garnet abrasive is iron aluminum silicate, $Fe_3Al_2Si_3O_{12}$. Garnets have unique and high performance compared to that of certainly available quartz.

Corundum and Emery Abrasives:

Corundum is naturally occurring mineral and it is nothing but aluminum oxide with high purity. Emery is a mixed oxides of aluminum oxide and iron oxide (Al_2O_3. Fe_2O_3). Corundum finds application as polishing of optical components. Emery has application of polishing metals with the coated products. Emery has hardness between quartz and corundum.

Diamond Abrasive:

Diamond was found first in India over 2000 years ago. It is a high quality abrasive and hence, it finds applications in the form of grain and powder in addition to single crystal abrasive.

Synthetic Abrasives:

Fused aluminum oxide
Fused Zirconia –aluminum oxide
Silicon Carbide
Boron Carbide
Diamond
Cubic Boron Nitride

Fused aluminum oxide abrasive:

As stated earlier section, abrasive performance of natural corundum and emery are quite different from each other. Therefore, it gives an insight to maximize the abrasive performance of fused aluminum oxide (synthetic aluminum oxide).

Bauxite ore is directly used to get fused aluminum oxide with relatively high aluminum oxide content by Bayer process. Thus, the final fused aluminum oxide (white fused Al_2O_3) that is regularly synthesized by Bayer process contains

Al_2O_3 = 82%
SiO_2=8%
Fe_2O_3=8%
TiO_2=4%.

However, indirect Bayer process yields higher aluminum oxide of 99%.

Fused Zirconia-aluminas Abrasives:

With the invention and advancement of fused zirconia-alumina abrasives outstanding commercial development took place in the last two decades in the area of abrasives. It has a higher efficiency of performance of abrasive when compared to that of fused alumina abrasive.

In the commercial production of fused zirconia-Al_2O_3 abrasives, about 25 wt.% to 40wt.% of zirconia is present in this type of abrasives. Again, alumina can be produced from Bauxite ore by the Bayer process and zirconia is from the Zircon mineral, where Zircon is a zirconium silicate. Both the alumina and zirconia are fused by electric-arc casting furnace with primary crystals of aluminum oxide in a matric of eutectic $Al_2O_3 – ZrO_2$ mixture.

In addition to higher efficiency of abrasive performance of $Al_2O_3 – ZrO_2$, they have a longer life on high speed and high pressure grinders.

Silicon carbide abrasive:

Silicon carbide was the first synthetic abrasive that was discovered by Edward G Acheson. When Acheson was attempting to synthesize diamond by an electric – arc heating process, some crystals were obtained, and Acheson though that the crystals were combination of carbon and Corundum, hence named as carborundum. Later, the crystals were determined to be silicon carbide, but the carborundum name is still retained for silicon carbide.

Abrasive silicon carbide is produced in an electric resistance furnace by the reaction between silica and coal as written by chemical equation below.

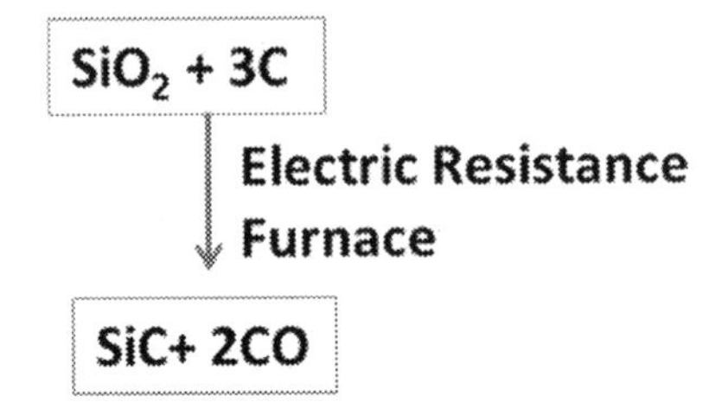

Several grades of silicon carbide abrasives are produced. Silicon carbide abrasives are used in grinding purpose of low tensile, nonferrous metals such as aluminum, brass, copper and nonmetallic such as glass, stone, concrete, ceramics and refractories.

Boron Carbide Abrasive:

Even though boron carbide abrasive is unsuited for use in bonded and coated products, it is mixed with SiC as abrasive.

Boron carbide, B_4C is produced in electric resistance furnace by a chemical reaction between boron oxide and coke according to the equation shown below.

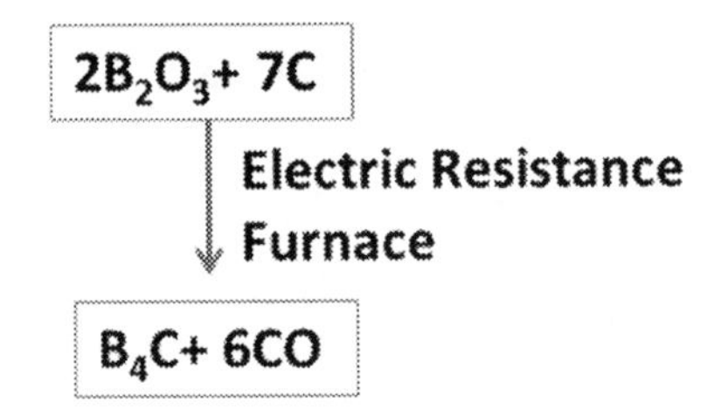

Diamond Abrasive:

Commercially synthetic diamond is produced under high pressure of 5 to 6.5 GPa and high temperature of greater than 1400°C. It is an important abrasive that was a breakthrough in the development of synthetic abrasive in the 20th century. Thus, it can eliminate shortage of natural diamond also and it can be possible to make various grade levels of diamond abrasives.

In general, synthetic diamond finds application for grinding, drilling and saving cemented carbides and a wide variety of plastics, glass, stone, concrete, ceramics, refractories, silicon and quartz.

Cubic Boron Nitride Abrasive:

Next to diamond, invention of boron nitride is the second and most important synthetic abrasive in the 20th century. Thus, cubic boron nitride was introduced commercially in 1969.

Very similar synthetic method adopted for the manufacture of diamond, high pressure and elevated temperature are prime conditions to make cubic boron nitride.

It has a good hardness (but inferior to diamond only), strength, low coefficient of friction during operation, thermal and chemical stability over 1000°C.

It is a good abrasive particularly for ferrous metals and hence, its production is getting increased for grinding the ferrous metals that have applications in the automobile industry.

Selective Question and Answer:

1. Write notes on solid lubricants:

Layered structure compounds such as graphite and molybdenum disulphide find application as solid lubricants. As a representative, the structure of graphite is given below in the Figure 2.

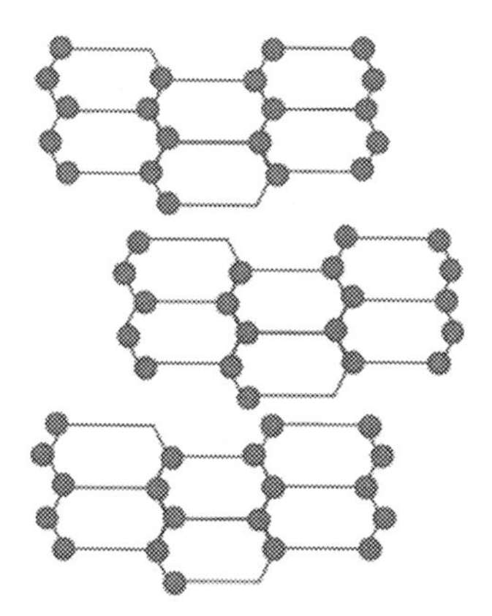

Fig.2: Structures of Graphite

Each layer of graphite consists of one atom thick of carbon chain with strong covalent bonding, and these layered are linked by weak force. In the case of MoS_2 layered compound each layer consists of MoS_6 octahedra units linked by edge sharing. These layers have again weak bonding. This layer structures favor for sliding without destruction and hence, these two layered compounds find application as solid lubricants.

Questions:

1. What is cement?
2. What are two steps in the production of cement?
3. What are the main constituents found in cement production?
4. How does tricalcium silicate differ from dicalcium silicate?
5. Describe chemistry of cement hydration
6. What is Ettringite phase?
7. What is induction period of hydration of cement?
8. Outline chemistry of post-induction period of hydration of cement
9. What is calcium silicate hydrate (CSH) gel formation?
10. Describe chemical corrosion of cement
11. How do marine environmental pollutions cause cement degradation?
12. What are five main components of Portland cement?
13. Describe chemistry of five main components of Portland cement.
14. What are refractories?
15. Describe three classifications of refractories with one example on each
16. What are basic refractories? Give a couple of examples
17. What are mixed oxide refractories and how do they differ from non-oxide refractories?
18. Concisely outline thermal properties of refractories
19. Briefly outline mechanical properties of refractories
20. Briefly outline thermal and electrical conductivities of refractories
21. Outline each step of manufacture of refractories
22. What are abrasives?
23. Outline classification of abrasives
24. Describe physical properties of abrasives
25. Give a few examples for natural abrasives and synthetic abrasives
26. What are fused ZrO_2-Al_2O_3 abrasives?
27. What is boron carbide abrasive?
28. What is cubic boron nitride abrasive?

Water Treatment

Objectives:

1. To emphasize the need for the chapter
2. To state major sources of water
3. To state water impurities
4. To define water treatment technologies
5. To illustrate the effect of minerals in water
6. To define hardness of water
7. To outline determination of hardness of water
8. To state consequences of minerals in water boiler
9. To illustrate various techniques of removal of minerals from water
10. To state differences among lime soda precipitation, zeolites separation and ion-exchange method
11. To state desalination process

Need:

Water is important and is needed in our daily lives and hence, quality of water should be improved and preserved due to not only increasing population but also increasing waste water from chemical and industrial products. In fact, 80% of earth surface is covered with sea water, presence of 3.5% NaCl salt prevents for human daily uses. Therefore, in order to explore available water sources and do recycle the available small water by cheap and clean technologies.

There are four major sources for water and these are

(1) Surface water (of flowing and static)
(2) Underground water (of wells)
(3) Sea water and
(4) Rain water.

Normally identified impurities in water are

(a) dissolved mineral matter: Ca, Mg hardness; Carbonates and bicarbonates alkalinity, mineral acidity and metal salts.
(b) Dissolved gases such as CO_2, O_2, N_2, H_2S etc.
(c) Turbidity due to inorganic and organic particles which clearly reduce clarity of the water
(d) Color and odor due to colloidal and small organic impurities &
(e) Micro organisms such as bacteria and virus.

For the water treatment due to presence of above said impurities from "A" to "E", several technologies are being explored. The category, classifications and summary of water treatment and recycling technologies are given in the Outline 1.

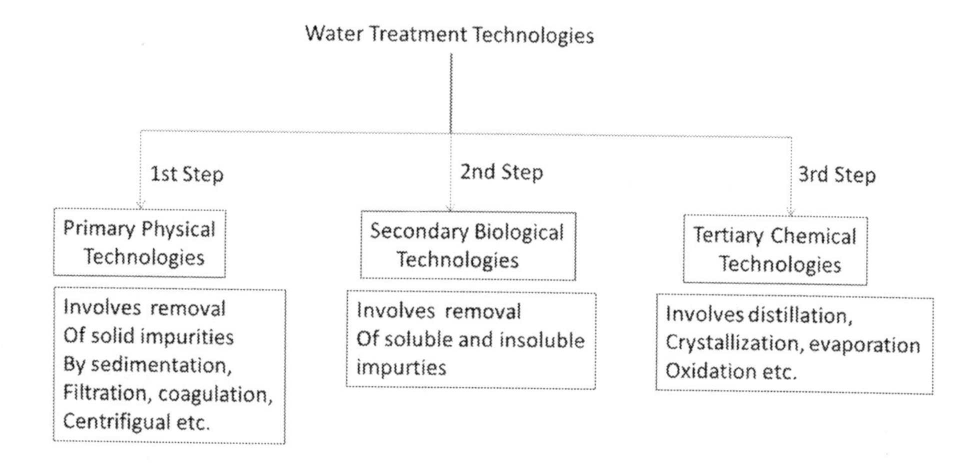

Outline 1: Summary of water treatment technologies

In the following section, what is the role of minerals in water?, How do the minerals affect the action of soap in water, How do you determine the minerals in water in terms of Hardness of water? What are the consequences of minerals in water boiler? What are the ways by which minerals can be removed from water? And finally, what are the disadvantages uses of minerals in drinking water? are the subjects of topic of interest.

Hardness of water:

Presence of dissolved minerals containing calcium and magnesium in natural water is called Hard water. Thus, carbonate hardness (or Temporary hardness) of water is due to presence of Calcium carbonate ($CaCO_3$), Magnesium carbonate ($MgCO_3$), Calcium bicarbonate [$Ca(HCO_3)_2$] and Magnesium bicarbonate [$Mg(HCO_3)_2$]. The non-carbonate hardness (Permanent hardness) of water is due to $CaSO_4$, $CaCl_2$, $MgSO_4$ and $MgCl_2$.

Estimation of Hardness of water:

Hardness of water is estimated in terms of equivalents of $CaCO_3$. Therefore, all the impurities causing hardness are converted to equivalent of $CaCO_3$. The formula/molecular weight of $CaCO_3$ is 100 g/mol

Thus, 100g $CaCO_3$ = 111g of $CaCl_2$=136g $CaSO_4$=95g of $MgCl_2$=120g of $MgSO_4$=162g of $Ca(HCO_3)_2$ = 146g of $Mg(HCO_3)_2$=164g of $Ca(NO_3)_2$ = 44g of CO_2

Soap Performance dependence on Hard Water:

Poor performance of soap on hard water is due to presence of mineral ions in hard water, which precipitates the sodium salts of fatty acids of soap. Thus, hardness of water is defined and determined by the soap performance dependence for the water. Thus, formation of rich, consistent foam is associated with satisfactory soap performance. Thus, the foam formation dictates the quality of hardness of water. If there is no foam formation, then, the water is hard water. If there is a rich foam formation, then, the water is called soft water. The unit for hardness of water is parts per million (1 ppm = 1 mg/liter).

Determination of hardness of water:

The hardness of water is usually determined by complexometric titration using complexing agents such as Ethylenediaminetetracetic acid (EDTA), Calmagite etc.

EDTA Method:

Principle:

EDTA is a strong complexing agent with most metal ions. Its structure is given below.

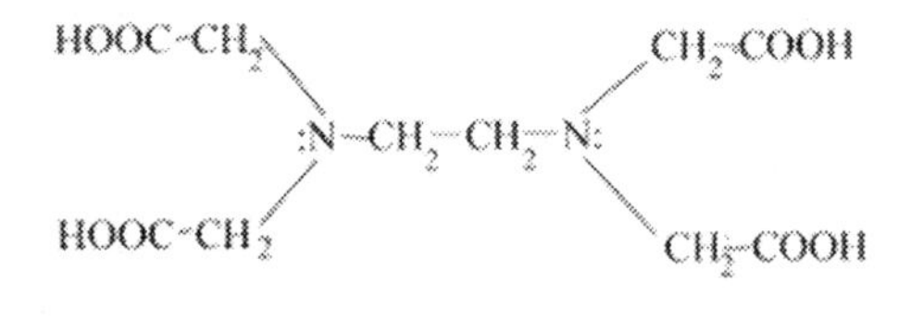

structure of EDTA

Disodium salt of EDTA is commonly used due to its high solubility in water and hence, its disodium salt of EDTA is denoted as Na_2H_2Y. Since hardness of water is due to Ca^{2+} and Mg^{2+} ions, Na_2H_2Y readily forms complex with Ca^{2+} and Mg^{2+} when hard water is titrated against standard solution of Na_2H_2Y using suitable metal ion indicator (example is Eriochrome Black T). As the

Na_2H_2Y titrant is added slowly to a hard water, Na_2H_2Y captures free Mg^{2+} and Ca^{2+} ions first. After all the Ca^{2+} and Mg^{2+} ions are consumed by Na_2H_2Y, Na_2H_2Y bounds to react with metal ion indicator to give bright color. The triumph of this complexometric titration to determine the hardness of water is due to complex formation constant of Mg^{2+} and Ca^{2+} with metal ion indicator is very low.

Experiment:

Hard water sample solution preparation:

250 mL of 1000 ppm Ca^{2+} ion is prepared by the following way. Aqueous slurry of 0.25 g of anhydrous $CaCO_3$ with 3 to 4 mL of deionized water is made first. To this aqueous slurry, add 6 M HCl drop by drop to dissolve $CaCO_3$ until the solution becomes clear. Finally, dilute to 250 mL with deionized water. The resulting concentration of the solution is about 250 mg $CaCO_3$ in 250mL (which is equal to 1000 ppm hard water).

pH 10 Buffer solution Preparation:

Dissolve 64g of NH_4Cl in 200 mL of distilled water. Add 570 mL of concentrated ammonia and dilute to 1 litre with distilled water. The pH=10 buffer solution is required to make the complexometric titration reaction instantaneously so that the outcome of the results will instant.

Calmagite Indicator Solution Preparation:

Dissolve 0.5 g of Calmagite in 1 litre of distilled water. Stir the solution for one hour. In spite of its long shelf life, the indicator should be prepared fresh every month.

EDTA Titrant solution preparation:

A solution of approximately 0.010 M of EDTA is prepared by weighting about 1 g of the disodium salt dehydrate of EDTA. Place this amount into a 500 mL flask and add 250 mL of deionized water to dissolve.

$Na_2MgEDTA$ Reagent solution preparation:

Weigh out 39.4 g of $Na_2MgEDTA$ and dissolve the reagent in 750 mL of deionized water. Add 15 mL of concentrated NH_4OH and 10 drops of

Calmagite indicator. Add small portion of Na_2EDTA salt until the solution is almost blue. Dilute this solution to 1000 mL using deionized water.

Procedure:

Take 25mL of hard water sample in a conical flask, to this solution add a few drops of pH=10 buffer solution. In addition to this, add about 10 drops of $Na_2MgEDTA$ reagent and a few drops of freshly prepared indicator. Now, the solution is ready to titrate against EDTA solution. The end point of titration should be color change from Wine red to blue. Repeat the titration two more times to get concordant value.

Consequences of minerals in water boiler:

Water boiler is used in industries to get steam. If hard water is used directly, then concentration of minerals in boiler increases as steam is produced. After having reached the saturation of mineral ions, they naturally tend to precipitate inside of the boiler and thus, the process leads to formation of so called scale and sludge formation.

Removal of Minerals in water:

The removal of minerals in water is essential and indeed is needed for industrial and domestic applications. Otherwise, in industry application, scale and sludge formation in boiler occurs. In domestic and household applications, hard water precipitates the soap solution. The process of removal of minerals in water is called softening of water. There are three conventional ways by which the softening of water can be achieved. The well-known and well-established methods are precipitation method (cold/hot lime-soda method), separation method (use of microporous zeolites) and ion-exchange method (using so called resins containing H^+/Na^+) that are very briefly outline below one by one.

Cold Lime-Soda Ash method:

This method was found back in 1841 by the Prof. Thomas Clark, Scottish professor of chemistry at Aberdeen University. Thus, softening of water is usually achieved by this method by addition of lime to precipitate the minerals as $CaCO_3$ and $Mg(OH)_2$. Lime contains $Ca(OH)_2$ and Soda ash contains $NaCO_3$, that are employed to precipitate [Ca^{2+}] and [Mg^{2+}] ions present in the hard water. Table 1 below summarizes the various possible reaction of softening of water by Lime-Soda method.

Table 1: Summary of Softening of water by Lime-Soda Method

Role of Reagents, $Ca(OH)_2$ and Na_2CO_3	**Reactions due to reagents**
$Ca(OH)_2$ removes dissolved CO_2 & H_2S	$Ca(OH)_2 + CO_2 \rightarrow CaCO_3(s) + H_2O$ $Ca(OH)_2 + H_2S \rightarrow CaS(s) + 2H_2O$
$Ca(OH)_2$ neutralizes HCl and H_2SO_4	$Ca(OH)_2 + 2HCl \rightarrow CaCl_2 + 2H_2O$ $Ca(OH)_2 + H_2SO_4 \rightarrow CaSO_4 + 2H_2O$
$Ca(OH)_2$ precipitates temporary hardness of $Ca(HCO_2)_2$ & $Mg(HCO_2)_2$	$Ca(OH)_2 + Ca(HCO_2)_2 \rightarrow 2CaCO_3(s) + 2H_2O$ $2Ca(OH)_2 + Mg(HCO_2)_2 \rightarrow Mg(OH)_2(s) + 2CaCO_3(s) + 2H_2O$
$Ca(OH)_2$ precipitates permanent hardness of $MgCl_2$ and $MgSO_4$	$Ca(OH)_2 + MgCl_2 \rightarrow Mg(OH)_2(s) + CaCl_2$ $Ca(OH)_2 + MgSO_4 \rightarrow Mg(OH)_2(s) + CaSO_4$
$Ca(OH)_2$ precipitates $FeSO_4$ & $Al_2(SO_4)_3$	$Ca(OH)_2 + FeSO_4 \rightarrow Fe(OH)_2(s)$, unstable $+ CaSO_4$ $2Fe(OH)_2 + H_2O + O \rightarrow 2Fe(OH)_3(s)$ $3Ca(OH)_2 + Al_2SO_4 \rightarrow 2Al(OH)_3(s) + 3CaSO_4$
Na_2CO_3 precipitates permanent hardness of $CaCl_2$ and $CaSO_4$	$Na_2CO_3 + CaCl_2 \rightarrow CaCO_3(s) + 2NaCl$ $Na_2CO_3 + CaSO_4 \rightarrow CaCO_3(s) + Na_2SO_4$

Since the process of precipitation of minerals at room temperature, it is usually termed as cold method. This method was later modified by the John Henderson porter using Soda ash in addition to lime. Using the principle of lime-soda ash method, there are four basic types of process known now. These are the sludge blanket, conventional, catalyst and intermittent or batch process. Very briefly, batch process is outlined here.

Batch Process of Cold lime-soda method:

The first and foremost step for engineering this method is the design and construction of reactor. The reactor consists of mechanical stirrer for homogeneous of precipitation of minerals as $Mg(OH)_2$ and $CaCO_3$ by adding reagent at the top center of the reactor while hard water is fed into the reactor at the top left corner of it. The precipitate is collected at the left bottom of the reactor and soft water is removed from the reactor at the right top side of the reactor. However, because of room temperature operation of it, it requires a couple of hours to completely settle down the precipitate from fine solid of mineral ions.

Hot lime-soda ash method:

In this method, unlike at room temperature, the removal of minerals is usually carried out at elevated temperature of boiling point of water. Hence, this method is called hot lime soda ash method. The increase in temperature accelerates the precipitation of minerals more significantly. Also, this process helps to grow the precipitates denser and heavier so that the separation of solid precipitate is much easier and quicker.

The design of hot lime soda ash method is a critical and important for this method. Since the temperature of operation is close to water boiling point of 100°C, additional inlet for steam is required to heat up the reaction and as it passes through the reaction media it forms water that facilitates the formation of dense precipitates and it is now quick to recover soft water in the outlet. Also, the precipitating reagent required is exact amount for the complete removal of minerals from hard water and thus, reagent waste can be minimized.

Ion exchange and zeolite methods:

Ion exchange method involves replacement of mineral ions such as Mg^{2+}, Ca^{2+} by Na^+, H^+ etc present in the solid resins.

Later, the capacity of removal of minerals was increased to large extent by use of microporous resins. The increment is due to higher specific surface area exposed to mineral water. This was known 100 years back. Now, Prof. J Thomas Way studied extensive work on ion exchange properties of zeolites, which is known as microporous materials. The zeolites now a days are widely and wisely used for separation of organic small molecules by column method. In addition to natural zeolites that find applications in water softening, plethora synthetic zeolites are also known with varying size and shape of pores.

There are four major types of ion exchange resins. These are strongly acidic and weakly acidic cation exchanges (two of them) and strongly basic and weekly basic anion exchangers (rest of two). Depending upon the presence of functional groups in the ion-exchanges, term strength or weak is referred. Cation exchange has either H^+ or monovalent alkali metal ion such as Na^+. For anion exchange, either OH^- or Cl^- is present with it. These ion exchangers can be regenerated by treating with acid for cation exchanger and base for anion exchanger. The ion-exchange reaction is usually represented as shown below.

$$Ca^{2+}/Mg^{2+} + H_2R \rightarrow CaR/MgR + 2H^+.$$

In the ion exchange process, the exchange of ions occurs continuously until there is H+ available for cation exchange.

A few properties of water softening are compared with the three methods of water softening (lime-soda precipitation method, zeolite ionic separation method and ion-exchange method) below in the table 2.

Table 2: Comparison of water softening by three different methods

Process of softening of water	Lime soda precipitation method	Zeolites separation method	Ion-exchange method
Hard to soft water	Good conversion	Better conversion	Best conversion
Suspended particle water treatment	The best method to separate particles	Good method to separate particles	Not good a good method to separate particles
Acidic water treatment	Can take of acidic water	Can't take care acidic water	Can help to remove acidic ions
Method of water treatment	Batch method is suitable	Continuous method is available	Continuous method is available
Hardness	Low hardness value can be obtained	Very Low hardness value can be reached	The lowest hardness value can be achieved

Desalination process by Reverse Osmosis method:

Due to external pressure applied at the solution side, usually larger than the osmotic pressure, solvent flows from the solution to solvent through semi-permeable membrane against the spontaneous flow. Thus, the reverse osmosis is useful to filter solute from solution and in fact, this method is used for desalination process to purify the sea water into drinking water. Fig.1 below is the representation of reverse osmosis phenomena.

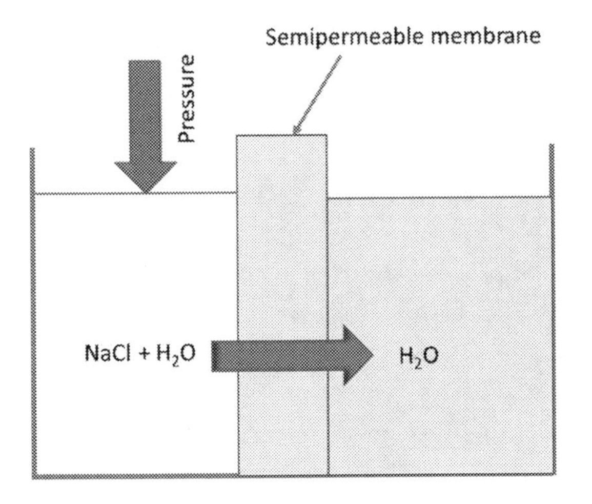

Fig.1: Outline of Reverse Osmosis Process for Desalination of water

Disadvantages of hard water:

The disadvantages of use of hard water in industrial applications and domestic applications are outlined below one by one.

Problems of hard water usage in industrial process:

There are several industry sectors that face problem of use of hard water. Some of them are textile industry, paper industry, sugar industry, dye industry and industry that uses boiler to make steam (please refer earlier section for scale and sludge formation in the boiler).

In the textile industry, yarn and/or cloth are usually washed with soap solution. If hard water is used during this process, function of soap is lost by precipitating as $Mg(OH)_2$ and $CaCO_3$ and hence, the tiny particles of precipitate attaches to cloth and/or yarn. Therefore, the process of washing is meaningless due to making the dirty of them.

In the paper industry, several chemicals are used in the processing. If hard water is used in the processing of paper, dissolved minerals present in the hard water might involve in the reaction with chemicals used in the paper processing. Then, it will lead to damage to paper and hence, whole processing of paper is meaningless.

Numerical Problems:

Problem 1:

Calculate the temporary, permanent and total hardness of a sample of water containing the following impurities.

$Mg(HCO_3)_2$ = 75 mg/L = 75ppm; $Ca(HCO_3)_2$ = 175 mg/L = 175ppm
$CaSO_4$ = 125 mg/L = 125ppm; $MgCl_2$ = 100 mg/L = 100ppm
$CaCl_2$ = 100 mg/L = 100ppm; NaCl = 200 mg/L = 200ppm

Given:

$Mg(HCO_3)_2$ = 75 mg/L = 75ppm; $Ca(HCO_3)_2$ = 175 mg/L = 175ppm
$CaSO_4$ = 125 mg/L = 125ppm; $MgCl_2$ = 100 mg/L = 100ppm
$CaCl_2$ = 100 mg/L = 100ppm; NaCl = 200 mg/L = 200ppm

Ask:

Temporary, permanent and total hardness.

Method:

$Ma(HCO_3)_2$ and $Ca(HCO_3)_2$ are responsible for temporary hardness $CaSO_4$, $MgCl_2$ and $CaCl_2$ are responsible for permanent hardness and NaCl does not contribute to hardness. It is noted that convert all the ions concentration with respect to $CaCO_3$.

Steps:

Temporary hardness	$CaCO_3$ Equivalents
$Mg(HCO_3)_2$ = 75 ppm	(100g/146g)*75ppm = 51.37 ppm
$Ca(HCO_3)_2$ = 175 ppm	(100g/162g)*175ppm = 108.02 ppm
Total of temporary hardness	159.39 ppm

Permanent hardness	
$CaSO_4$=125ppm	(100g/136g)*125ppm=91.911 ppm
$MgCl_2$ = 100 ppm	(100g/95g)*100ppm = 105.26 ppm
$CaCl_2$ = 100ppm	(100g/111g)*100ppm = 90.09 ppm
Total of permanent hardness	287.2613 ppm
Total hardness of water	Temporary + permanent hardness=446.651 ppm

Results:

Temporary hardness of water = 159.39 ppm; permanent hardness of water = 287.2613 ppm and total hardness of water = 446.651 ppm

Problem 2:

Calculate the total amount of lime (100% pure) and soda (100% pure) required for softening of 1 million liter of hard water that has the following chemicals responsible for hardness.

$Ca(HCO_3)_2$ = 50 mg/L; $Mg(HCO_3)_2$ = 75 mg/L; $MgSO_4$ = 75 mg/L; $CaSO_4$ = 50 mg/L; $CaCl_2$ = 25 mg/L and NaCl = 25 mg/L

Given:

$Ca(HCO_3)_2$ = 50 mg/L; $Mg(HCO_3)_2$ = 75 mg/L; $MgSO_4$ = 75 mg/L; $CaSO_4$ = 50 mg/L; $CaCl_2$ = 25 mg/L and NaCl = 25 mg/L

Ask:

Calculate the amount of soda and lime required for softening of 1 million liter of hard water.

Method:

First convert all the constituents responsible for hardening of water. Lime, $Ca(OH)_2$ takes care of converting $Ca(HCO_3)_2$, $Mg(HCO_3)_2$ and $MgSO_4$ into $CaCO_3$, $CaCO_3$ and $CaSO_4$. Soda takes care of converting $CaSO_4$, $CaCl_2$ and $CaSO_4$ formed by the lime process into $CaCO_3$.

Steps:

Salt	$CaCO_3$ equivalent
$Ca(HCO_3)_2$ = 50 ppm	(100g/162g)*50ppm = 30.86 ppm
$Mg(HCO_3)_2$ = 75 ppm	(100g/146g)*75ppm = 51.37 ppm
$MgSO_4$ = 75 ppm	(100g/120g)*75ppm = 62.5 ppm
$CaSO_4$ = 50 ppm	(100g/136g)*50ppm = 36.76 ppm
$CaCl_2$ = 25 ppm	(100g/111g)*25ppm = 22.52 ppm

$1Ca(OH)_2$ converts $1Ca(HCO_3)_2$ into $CaCO_3$; $2Ca(OH)_2$ converts $Mg(HCO_3)_2$ into $CaCO_3$; $1Ca(OH)_2$ converts $MgSO_4$ into $CaSO_4$;

As 100 ppm $CaCO_3$ = 74 ppm $Ca(OH)_2$

(74 ppm/100 ppm)[30.86+2(51.37)+62.5] = 145.114 ppm = 145.114 mg/L of $Ca(OH)_2$ is required by the role of lime.

145.114 mg of $Ca(OH)_2$ (lime) is required for 1 liter of water to soften. Therefore, 1 million liter of water require is that

145.114 mg x 1000000 = 145.114 Kg

$1Na_2CO_3$ converts $1CaCl_2$ into $CaCO_3$; $1Na_2CO_3$ converts $1CaSO_4$ into $CaCO_3$; and $1Na_2CO_3$ converts $1CaSO_4$ into $CaCO_3$;

As 100 ppm $CaCO_3$ = 106 ppm of Na_2CO_3

(106 ppm/100 ppm)[22.52+36.76+62.5] = 129.09 ppm = 129.09 mg/L of Na_2CO_3 is required by the role of soda.

129.09 mg of Na_2CO_3 (soda) is required for 1 litre of water to soften. Therefore, 1 million liter of water require is that

129.09 mg x 1000000 = 129.09 Kg

Results:

Lime required to soften the 1 million of water = 145.114 Kg and soda required to soften the 1 million of water = 129.09 Kg.

Problem 3:

Amount of lime required for softening of 1 liter of water is 333 mg. temporary hardness of water is 220 ppm. Permanent magnesium hardness is 10 ppm. What is the metal bicarbonate causing the temporary hardness?

Given:

Lime needed for 1 liter water softening = 333 mg; Temporary hardness = 220 ppm and permanent magnesium hardness = 10 ppm

Ask;

What is the metal bicarbonate type causing the temporary hardness?

Method:

$1Ca(OH)_2$ converts $1Ca(HCO_3)_2$ into $CaCO_3$
$2Ca(OH)_2$ converts $1Mg(HCO_3)_2$ into $CaCO_3$
$1Ca(OH)_2$ converts $1MgCl_2/MgSO_4$ into $Mg(OH)_2$.

Steps:

As 100 g $CaCO_3$ = 74 g $Ca(OH)_2$

If $Ca(HCO_3)_2$ is responsible for temporary hardness, then lime required is

= (74/100)(220+10) = 170.2 mg/L

If $Mg(HCO_3)_2$ is responsible for temporary hardness, then, lime required is

=(74/100)(220x2+10) = 333 mg/L

Result:

Sine lime amount is given as 333 mg/L, $Mg(HCO_3)_2$ is the metal bicarbonate that causes the temporary hardness.

Problem 4:

Calculate the amount of lime (100%) and soda (100%) required to soften 100 liter of water per day for the whole march month containing the following. $CaCO_3$ = 2.85 mg/L; $CaSO_4$ =50 mg/L; $MgCO_3$ = 0.52 mg/L; $MgCl_2$ = 80 mg/L; $MgSO_4$ = 100 mg/L; SiO_2 = 30 mg/L.

Given:

$CaCO_3$ = 2.85 mg/L; $CaSO_4$ =50 mg/L; $MgCO_3$ = 0.52 mg/L; $MgCl_2$ = 80 mg/L; $MgSO_4$ = 100 mg/L; SiO_2 = 30 mg/L

Ask:

Amount of lime and soda required for 100 liter per day for march month.

Method:

Convert all the salt into $CaCO_3$ equivalent.

Na_2CO_3 (Soda) reacts with $CaSO_4$ and converts it into $CaCO_3$

$Ca(OH)_2$ (lime) reacts with $MgCO_3$, $MgCl_2$, $MgSO_4$ and SiO_2 and convert them into $Mg(OH)_2$ and $CaSiO_3$ respectively.

Steps:

Salt amount	**$CaCO_3$ equivalent**
$CaCO_3$ = 2.85 mg/L	2.85 mg/L
$CaSO_4$ = 50 mg/L	(100/95)x50 = 52.631 mg/L
$MgCO_3$ = 50 mg/L	(100/84.314)x50 = 59.30 mg/L
$MgCl_2$ = 80 mg/L	(100/120)x80 = 66.66 mg/L
$MgSO_4$ = 100 mg/L	(100/162)x100 = 61.728 mg/L
SiO_2 = 30 mg/L	(100/60.08)x30 = 49.93 mg/L

As 100 g $CaCO_3$ = 74 g $Ca(OH)_2$

The lime required is

=(74/100)x[49.13+61.728+66.66+59.03] = 175.245 mg/L

For 100 liter water softening, the lime required is

= 175.245 mg x 100 = 17524.5 mg = 17.5245 g per day

For March month, 31 days, the lime required is

= 17.5245 x 31 = 543.259 g

As 100 g $CaCO_3$ = 106 g Na_2CO_3

The soda required is

= (106/100)x[52.631] =55.789 mg/L

For 100 liter softening of water, the soda required is

= 55.789 x 100 = 5578.9 mg per day

For March month, 31 days, the soda required is

= 5578.9 x 31 =172.945 g

Results:

The lime required is =543.259 g
The soda required is = 172.945 g.

Problems:

1. What are the four major sources of water?
2. What are impurities found in natural water?

3. What is water treatment?
4. What are water treatment technologies known?
5. What is hardness of water?
6. How do you estimate hardness of water?
7. How do you determine hardness of water?
8. How do you remove minerals?
9. What is cold lime-soda method?
10. What are ion-exchange and zeolite methods?
11. What is reverse osmosis (RO)?

Environmental Pollution

Objectives:

1. To classify the environmental pollution
2. To elaborate air pollution such as global warming, skin cancer and acid raid
3. To elaborate depletion of ozone
4. To elaborate acid rain
5. To outline water pollution'
6. To outline types of water pollutants
7. To explain oxygen demand and water pollution indicators
8. To explain chemical oxygen demand (COD)
9. To outline biological oxygen demand (BOD).

Environmental pollution deals with air, water and soil pollutions. Chemistry and causes for these air, water and soil pollutions (Outline 1) are the subject of this chapter. The main reason for these pollutions at the earth surface is due to human activities.

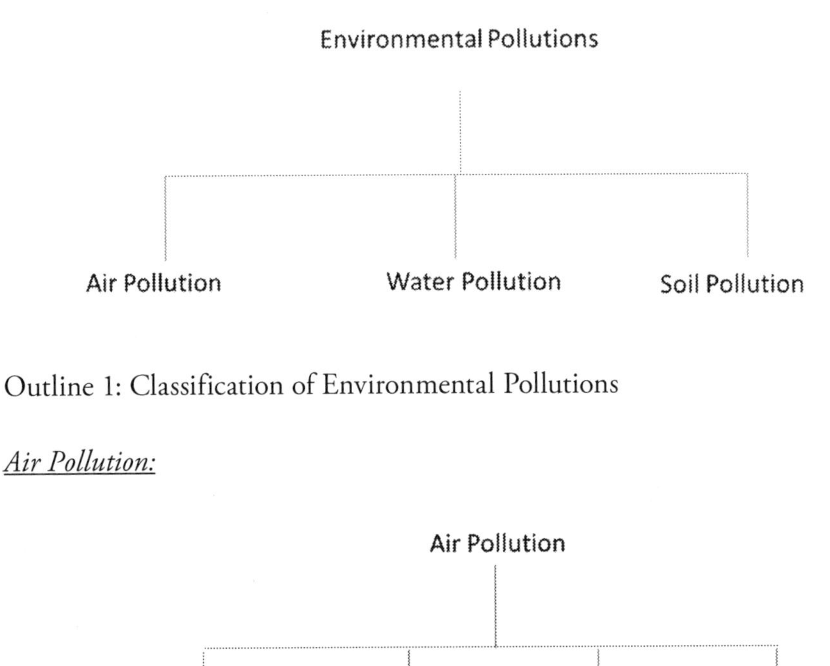

Outline 1: Classification of Environmental Pollutions

Air Pollution:

Outline 2: Topics covered in the classification of Air Pollution

Inorganic gases, organic gases and lightweight particles are responsible for the air pollutions. Some examples for inorganic gases that cause air pollution are NOx, SO_2, CO, CO_2, H_2S, NH_3, Cl_2 etc. Some examples for organic gases are CH_3CHO, HCHO, Chlorinated hydrocarbons. Some light weight particles are dust, carbon, ash, asbestos etc. These air pollutions are due to combustion of fuels and gases into CO_2, unburned hydrocarbons, NO_x etc. These air pollutions lead to major causes such as greenhouse effect & global warming, ozone depletion & skin cancer and acid rain (Outline 2)

Greenhouse Effect:

The greenhouse effect ultimately increases the temperature of earth surface. The gases responsible for greenhouse effect, the greenhouse effect, the effect of greenhouse effect on the temperature of the earth surface and finally, how can be the greenhouse effect avoided are described meaningful way in this topic of interest.

Earth absorbs energy (UV/Vis/IR) from the sun and it also emits energy. Its emission energy is mainly composed of Infrared region of electromagnetic spectrum. The IR emission from the arch can be absorbed by air molecules such as CO2 and H2O molecules and remits as heat to the earth surface. Thus, earth surface temperature remains around 300K (Fig.1).

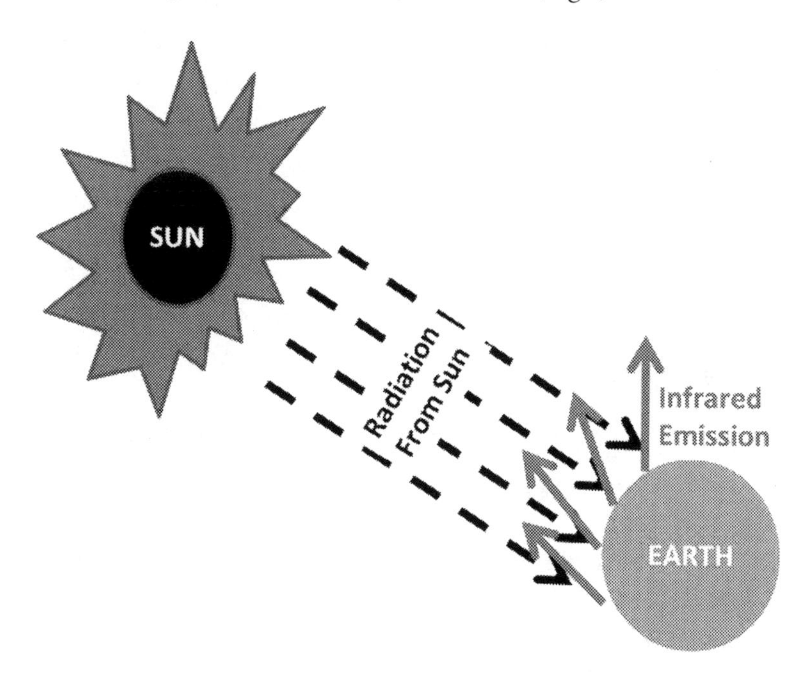

Fig.1: Solar absorption and Infrared emission by the EARTH.

Because of increase of trace amount of gases that absorb infrared radiation at the earth surface leads to enhanced greenhouse effect. Thus, greenhouse effect leads to obvious increase in temperature and hence, it is called artificial global warming.

The main reason for H_2O and CO_2 molecules that absorb infrared energy and emit it as heat energy is due to molecular vibrations excitation from their ground state to the excited states. Asymmetic or antisymmetric stretching of CO_2 molecule absorbs the infrared radiation. Similarly, bending vibrations of CO_2 can absorb Infrared radiation. However, CO_2 molecules do not absorb infrared radiation by symmetric stretching vibrations.

One strong absorption lies at 15 μm, due to bending vibration of CO_2. Another strong absorption at 4.26 μm is due to asymmetric stretching.

Other greenhouse gases are H_2O, Methane, Nitrous oxide (N_2O). The order of greenhouse gases responsible for global warming is $CO_2 > CH_4 > O_3 >$ CFCs $> N_2O$.

In order to keep the earth surface temperature unaffected, greenhouse gases should not be introduced more and more to the earth surface. Thus, combustion of coal, fuel and gasoline should be reduced. Harnessing full spectrum of electromagnetic radiation of solar energy is indeed essential and necessary for future energy resource through photovoltaics cells and/or solar cells. This will in future ultimately reduce the greenhouse gases greatly and save the earth.

Function, production & destruction and Depletion of Ozone:

Atmospheric oxygen (O_2) and ozone (O_3) molecules filter the sulight's UV components. Thus, O_3 molecule is responsible for most of the UV radiation from 220 – 320 nm (UV-B) due to its absorption. Whereas O_2 molecule filters UV from 120 to 220 nm radiation.

However, 1% reduction of Ozone level will result in allowance of 2% intensity of UV-B to earth surface. The UV-B portion of UV radiation can be absorbed by DNA molecules and will result in damaging reactions. The most skin cancers in humans are due to over exposure of UV-B from SUN light.

The Chapman mechanism, as described below (Scheme 1) is responsible for ozone production and destruction process in stratosphere.

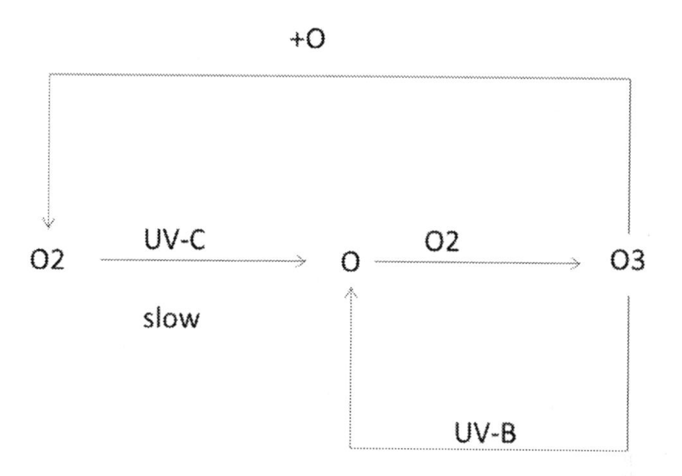

Scheme 1: Chapman Mechanism of Ozone production and destruction mechanism

At the stratosphere O_3 concentration is in the order of 10 ppm, which is greatly sufficient enough for UV-B and UV-C filtration from the SUN.

Major catalytic destruction of ozone in stratosphere is by halogenated hydrocarbons. This reduction in ozone concentration leaves a small opening for the major UV radiation from the SUN to reach earth surface. This small opening in ozone layer is called ozone depletion or hole. The major radical mechanism that creates ozone hole is shown below in the equations.

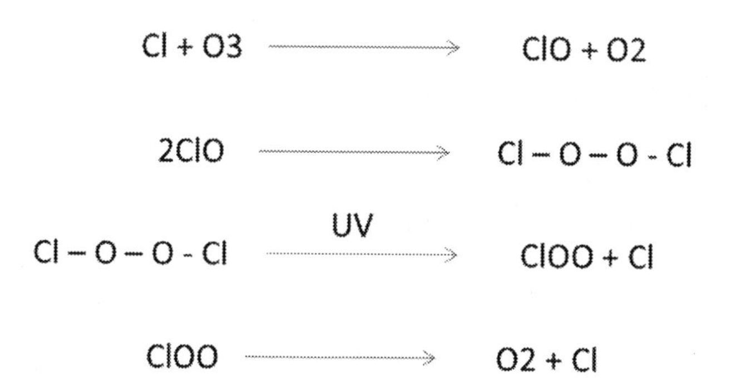

Chlorofluorocarbons (CFCs) are used in coolants and refrigerators due to nontoxic, nonflammable and nonreactive and useful condensation properties. These gases are also responsible for decrease in concentration of ozone layer.

Photochemical Smog:

Above certain level of ozone at the ground level leads to formation of smog by photochemical reaction is called photochemical smog.

Acid rain:

When water and carbon dioxide reacts reversibly to form weak acid of H_2CO_3 (carbonic acid), which in turn release proton, is the first observation for responsible for acid rain. The significant of acids in acid rain are H_2SO_4 (sulfuric acid) and HNO_3 (Nitric acid), formed from primary pollutants, namely sulfur dioxide, SO_2 and nitric oxide respectively, as shown below in the two equations.

$$SO2 + O2 \xrightarrow{H2O} H2SO4$$

$$NOx + O2 \xrightarrow{H2O} HNO3$$

Water Pollution:

Sources of water pollution:

There are three main sources for water pollution. They are domestic effluent waste, industrial effluent waste and agriculture waste. These three wastes essentially contribute to decrease the dissolved oxygen, to increase pH of water and to increase the toxic levels of water.

Types of water pollutants:

The main pollutants that contaminate water are organic waste (both biodegradable and nonbiodegradable), turbidity by suspended solids, N, S and P compounds from domestic, industrial and agriculture waste, toxic metals from mining, vehicle exhaust emission, pathogens from domestic waste etc. Table 2 below summarizes types of water pollutants and source of them.

Table 1: Various sources of water pollutants

Pollutant	**Source**
Organic waste	Industrial textile waste
N, S & P compounds	Agriculture waste
Thermal	Industrial cooling process
Toxic metals	Purification of metals, mining, irrigation with waste water, vehicle exhaust emission
Pathogens	Domestic waste water
Carcinogens	Industrial and Agriculture waste
Radioactive	Release from nuclear power plant

Organic pollutants can decrease the dissolved oxygen in water and hence, it leads to oxygen sag. Over a time, O_2 from atmosphere can dissolve the water to be replenished.

Oxygen Demand and Water Pollution Indicators

Dissolved oxygen in water is essential and it is constantly consumed by microorganism in water. If organic matter is discharged into water, the, dissolved O_2 is also consumed by oxidation of organic matter and hence, water demands oxygen from atmosphere. Thus, water quality is affected if organic wastes present in water. The dissolved O_2 level is also low in country like India due to solubility of O_2 gas decreases with increase in temperature. Thus, the dissolved oxygen amount in unpolluted water at 0°C is 14.7 ppm where as it is only at 7 ppm at 35°C. At 25°C, it is 8.7 ppm. Therefore, the un-polluted water should have dissolved O_2 in the order of 7±1 ppm.

The type and category of water pollution indicators are summarized in table 2.

Table 2: Summary of Water Pollution indicators

Type	**Category**
Physical	Temperature
	Turbidity
	Total suspended solids
Chemical	pH
	pE scale
	Dissolved O_2
	Nitrates
	Chemical Oxygen Demand (COD)
	Total Organic Carbon (TOC)
	Biochemical Oxygen Demand (BOD)
	Pesticides
	Metals
Biological	EPT index
	Trent biotic index

In this section, chemical type pollution indicators are explained one by one.

pH indicator:

If nitrogen, sulfur and phosphor containing compounds are dissolved in water, oxidation of these contaminated compounds results in formation of nitric acid, sulfuric acid and phosphoric acid. Therefore, not only dissolved oxygen is consumed but also pH of water decreases into acidic value. Thus, water quality can be easily correlated to pH value.

Dissolved O_2:

If organic compounds are dissolved in water, O_2 gas dissolved in water is consumed for the oxidation of organic compounds. Therefore, dissolved O_2 value in water can be correlated to water quality.

pE Scale:

Low pE value of water indicates the strong reducing power of dissolved substances in water. High pE value signifies the strong oxidation power of dissolved substances in water. Thus, pE is defined as the negative base – 10 logarithm of the effective concentration – the activity – of electrons in water.

Nitrates:

Bacteria can reduce some nitrate into nitrite ion in the baby as well as adult stomach as shown below.

$$NO_3^- + 2H^+ + 2e^- \rightarrow NO_2^- + H_2O$$

This nitrite oxidizes in combination with Hemoglobin in blood and thus, O_2 uptake by hemoglobin is reduced. Therefore, excess nitrate ion in drinking water is a potential health hazard. Sometime presence of nitroamines in food and water increases stomach cancer. Also, N-nitroamines are carcinogenic in animals.

Chemical Oxygen Demand (COD):

This is another water pollutant index or indicator, which is the measure of pollutants in water. Thus, oxygen demand in water can be easily evaluated by chemical oxygen demand. For COD test, powerful oxidant than O_2, for example acidified potassium dichromate is used to completely oxidize the organic waste present in water. The remaining potassium dichromate is back titrated with ferrous ion to the end point. The oxidation of organic matter by $K_2Cr_2O_7$ is written as

$$Cr_2O_7^{2-} + 14H^+ + 6e^- \rightarrow 2Cr^{3+} + 7H_2O$$

Whereas, oxidation of organic molecules by O_2 molecules is written as

$$O_2 + 4H^+ + 4e^- \rightarrow 2H_2O.$$

Therefore, 1 mole of potassium dichromate is equivalent to 6 electrons/4 electrons = 1.5 moles of O_2.

Thus, amount of potassium dichromate consumed can be correlated to oxygen consumed by organic compounds present in water, which in turn dictates the remaining dissolved oxygen.

Total Organic Carbon (TOC):

Total organic carbon is yet another pollutant indicator. Thus, all the organic compounds dissolved in water are oxidized into CO_2 by photocatalysis. Then, the amount of CO_2 evolved is evaluated either by Gas Chromatography. The procedure employed for the complete conversion of organic compounds into CO_2 by photocatalysis is carried out as represented by the equation below.

Organic Compounds in water + UV radiation in presence of potassium peroxydisulphate

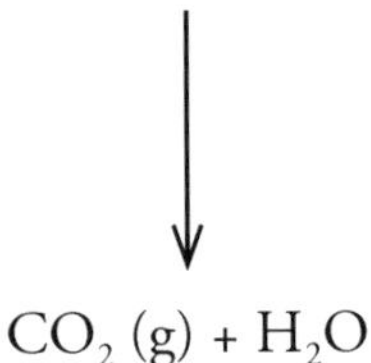

CO_2 (g) + H_2O

Thus, formation of amount of CO_2 is indicative of the presence of organic wastes in the polluted water.

Biochemical Oxygen Demand (BOD):

Biochemical Oxygen Demand represents the amount of dissolved O_2 consumed by oxidation of organic matters present in water. The oxidation of organic matters in water is catalyzed by microorganism present in water and hence, it is called as Biochemical oxygen demand. However, it requires longer time to estimate BOD when it is compared with that of Chemical Oxygen Demand (COD).

Ground Water Contaminants by Organic Chemicals:

The major problem with ground water is the contamination of organic chemicals. Thus, chlorinated organic solvents are the most prevalent pollutants in the round water. The examples include Trichloroethene (TCE) and perchloroethene (PCE). Moreover, ground water is the ultimate sink for organic contaminants.

Contaminants of soil by heavy metals:

Heavy metals such as Hg, Pd, Cd and As are more concern pollutants. They can combine and react with thiol group (-SH) of enzymes and hence, enzymetic control of metabolic reaction in human body is affected. Hg is considered as a major threat to health hazard element since it can be easily oxidized by bacteria. Then, Hg^{2+} is a reactive ion to destroy some useful micro organisms. Therefore, intense research activities are carried out to isolate the heavy metals present in water by microporous materials.

Questions:

1. Define environmental pollutions
2. What is air pollution?
3. What is acid rain?
4. What is greenhouse effect?
5. What are the main causes for ozone depletion?
6. Define COD and BOD.
7. What is TOC?

Chemistry of Explosives

Objectives:

1. To classify explosives
2. To define physical, chemical, and nuclear explosives
3. To outline chemical explosives
4. To outline classification of chemical explosives
5. To elaborate chemistry of primary explosives such as lead azide, mercury fulminate, lead styphnate, silver azide, and tetrazene
6. To outline secondary explosives such as nitroglycerine, nitrocellulose, picric acid, tetryl, TNT, Nitroguanidine, PETN, RDX, HMX, TATB, HNS, NTO, and TNAZ
7. To elaborate thermochemistry of explosives such as oxygen balance and estimation of fuel valence
8. To study decomposition products of explosives
9. To explain Kistiakowsky-Wilson (K-W) rules
10. To understand application of K-W rules
11. To explain Springall Roberts Rules
12. To study applications of Springall Robert Rules
13. To elaborate heat of reactions.

This chapter begins with the classification of explosion. Thus, explosion is divided into three groups based upon the mechanism by which explosion occurs.

1. Physical explosion due to high pressure;
2. Chemical explosion due to sudden and rapid chemical reaction;
3. Atomic or nuclear explosion due to nuclear chain reaction.

Among the three groups of explosion, the chemical explosion type attracts the most attention in the development of explosive materials while atomic explosion needs great care due to involvement of nuclear chain reactions. These three types of explosions are outlined below.

Physical Explosion:

A physical transformation is responsible for physical explosions. During a physical explosion, potential energy is converted into kinetic energy and a rapid raise in temperature is observed.

Volcano due to eruption is a typical example of a physical explosion, which occurs due to development of pressure.

Chemical Explosion:

A chemical reaction is responsible for chemical explosions and this leads to the evolution of lots of gases, accompanied by heat and light. Therefore, a chemical explosion is called an exothermic reaction. The chemicals responsible for chemical explosions are called chemical explosives.

Atomic or Nuclear Explosion:

An atomic explosion is due to nuclear chain reaction. Energy released is at a maximum in the atomic explosion, which is of course magnitude higher than that of chemical explosion. During an atomic explosion, heavy flux neutrons, gamma rays, and ultraviolet and infrared radiation are produced. While gamma rays are harmful to human, neutrons released in the atomic explosion lead to human fatal. Outline 1 below summarizes the classification of explosive materials.

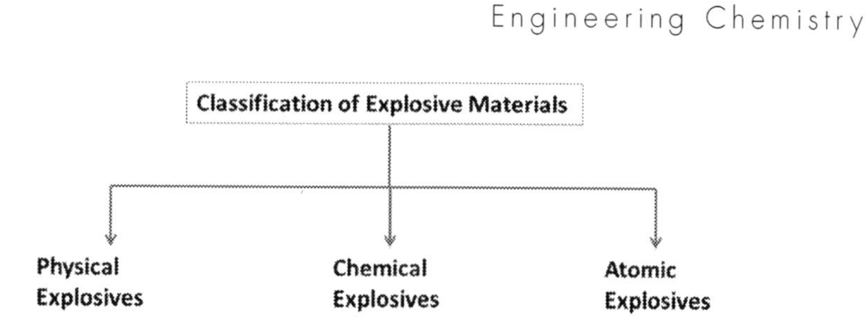

Outline 1: Classification of Explosives

The following section is mainly devoted to chemical explosives. Concern over classification of chemical explosives is based upon (i) the nature of explosives, (ii) the performance of explosives, and (iii) the application of explosives. Firstly, classification of chemical explosives based upon nature of functional group present in it is outlined. Properties of explosives are highly dependent on the nature of functional groups of chemical explosives, thus correlation between properties of explosives and structure of chemical explosives can be possibly understood. Table 1 below summarizes the groups present in chemical explosives.

Table 1: Summary of chemical explosives

Group responsible for explosion	Explosive compounds
Peroxide and ozonide group (example: - O – O -; - O – O – O -)	Peroxide and ozonide can be inorganic and organic compounds
Chlorate and Perchlorate group (example: - $OClO_2$ and – $OClO_3$)	Chlorate and perchlorate can be inorganic and organic compounds
Bond between Nitrogen and halogen (example: - N – X2 where X = halogen)	Compounds in which existence of bond between Nitrogen and Halogen
Nitrite and Nitrate group (example: NO_2 and $–ONO_2$)	Compounds can be inorganic and organic substances
Azide group (example: - N=N- and –N=N=N-)	Compounds can be inorganic and organic substances

Fulminate (example: -N=C)	Fulminate
Acetylne group (example: -C=C-)	Compounds can be organic and metal acetylates
Organometallic linkage (example: M-C)	Compounds can be organometallic compounds with Metal – carbon bond

Chemical explosives can be classified into three groups based upon their performance and uses: primary explosives, secondary explosives, and propellants as shown in outline 2 below.

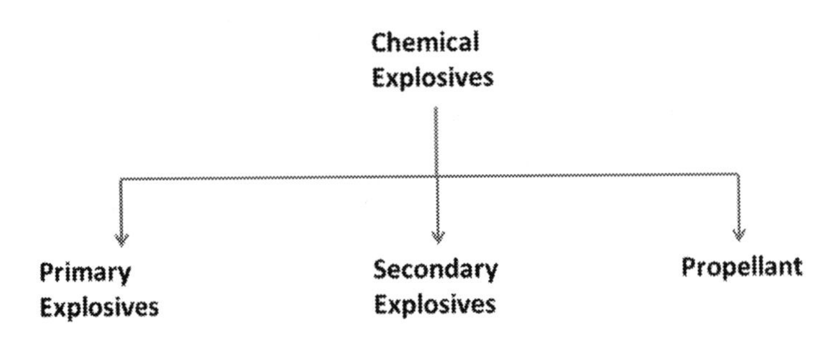

Outline 2: Classification of chemical explosives based upon their performance and uses.

Outline 3 below outlines classification of explosive materials and their subject of interest in this chapter.

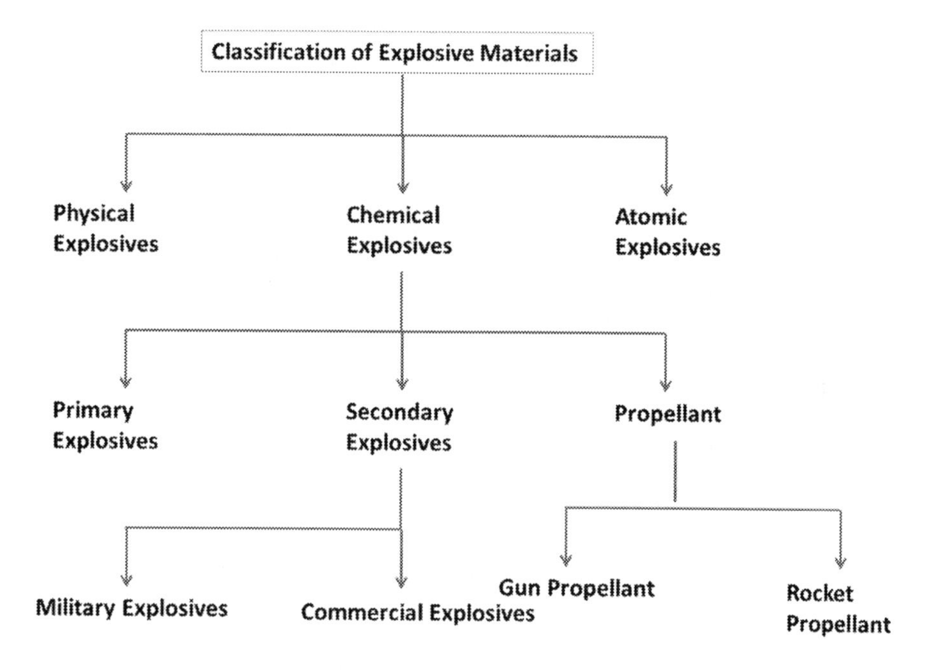

Outline 3: Primary explosives, secondary explosives, and propellants are discussed below one by one.

Primary Explosives:

Primary explosives are high explosives and can detonate when they are subjected to heat or shock. During detonation, the molecules undergo rapid and instant decomposition accompanying with production of tremendous amount of heat and/or shock. A typical example for primary explosives is lead azide (PbN_6). Its explosive chemical reactions are shown below.

$$\frac{1}{2} PbN_6 \longrightarrow \frac{1}{2} Pb^{2+} + N_3^- \longrightarrow \frac{1}{2} Pb + N_2 + N$$

Lead azide decomposes into a stable element, Pb and molecule, N_2 and nitrogen atom. This reaction is endothermic reaction and it requires 213 kJ of energy. Since the nitrogen atom is not in a stable situation, it further reacts with PbN_6 (lead azide) into stale lead and the nitrogen molecule as shown below.

$$\frac{1}{2} PbN_6 + N \longrightarrow \frac{1}{2} Pb^{2+} + N_3^- + N \longrightarrow \frac{1}{2} Pb + 2N_2$$

However, the above reaction is a highly exothermic reaction and it produces 657 KJ of energy. The lead azide decomposes into lead, Pb, and nitrogen molecule, N_2. The lead and nitrogen molecule are a stable element and molecule, respectively.

Other examples for primary explosives are lead styphnate (trinitroresorcinate), lead mononitroresorcinate (LMNR), potassium dinitrobenzofurozan (KDNBF), and barium styphnate. The chemistry of primary explosives is outlined along with their structures below one by one.

Mercury Fulminate:

Structure and Synthesis:

It is prepared by treating a solution of mercuric nitrate with alcohol in nitric acid. Its structure is shown below.

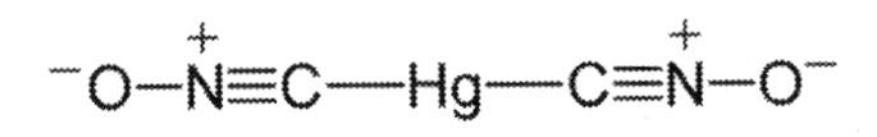

It decomposes with explosive nature into stable products as shown below with a chemical reaction. The stable products are CO, N_2, and Hg.

$Hg(CNO)_2 \rightarrow 2\ CO + N_2 + Hg$

Properties:

It is very sensitive to sunlight and it decomposes into stable products with the evolution of gas.

Lead Azide:

Structure and Synthesis:

Curtius prepared lead azide for the first time in 1891. Its preparation involves addition of lead acetate to a solution of sodium or ammonium azide. Its structure is shown below.

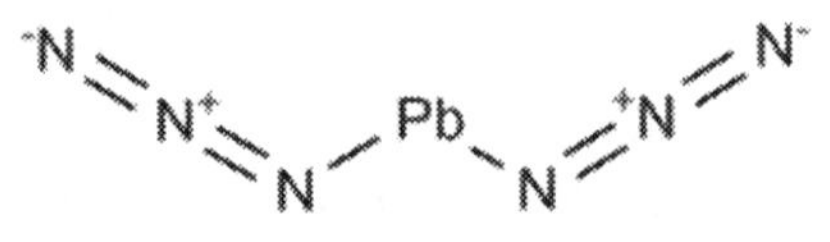

It has two allotropic forms. They are the more stable alpha form with orthorhombic and the beta form is monoclinic. Preparation of the alpha form involves a rapid method whereas the beta form synthesis involves a slow diffusion method.

Properties:

Shelf life of lead azide is usually good in dry conditions. But it decomposes readily on exposure to moisture, oxidizing agents, and ammonia.

Uses:

It is widely used in detonators because of its high capacity to initiate other secondary explosives to detonation.

Lead Styphnate:

Structure and Synthesis:

It is usually prepared by adding a solution of lead nitrate to magnesium styphnate. It is also known as lead 2,4,6 – trinitroresorcinate with a molecular formula of $C_6H_3N_3O_9Pb$. Its structure is shown below.

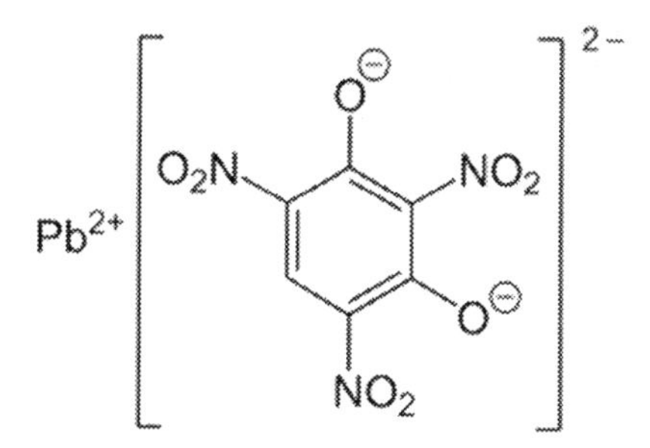

Properties:

Lead styphnate is insoluble in water and most organic solvents. It is non-hygroscopic. It is stable at room temperature and high temperatures up to 75°C. It is exceptionally resistant to nuclear radiation. It is very sensitive to flame and electric spark. It is a primary explosive due to its high metal content (44.5%).

Uses:

It is used in ignition caps and in the ASA where A = lead azide, S = lead styphnate, and A=aluminum. This mixture can be detonators.

Silver Azide:

Structure and synthesis:

Its synthesis is very similar to that of lead azide synthesis and thus its preparation involves the action of sodium azide on silver nitrate in an aqueous solution. Its structure is shown below.

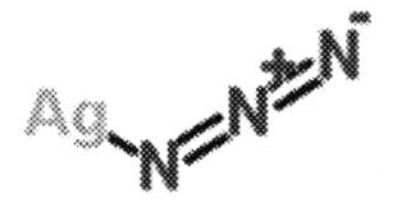

Properties:

It is slightly hygroscopic and very vigorous initiator. Like lead azide, silver azide decomposes on exposure to ultra-violet radiation. Its photochemical decomposition depends upon the intensity of light.

Tetrazene:

Synthesis and structure:

Preparation of Tetrazene or tetrazolyl guanyltetrazene hydrate ($C_2H_8N_{10}O$) was reported for the first time by Hoffmann and Roth in 1910 and involves the action of a neutral solution of sodium nitrite on aminoguanidine salts. Its structure is shown below.

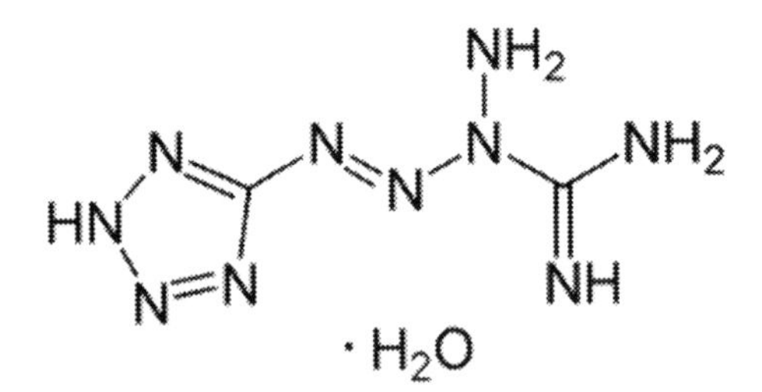

Properties:

Tetrazene is slightly hygroscopic but stable at ambient temperatures. Evolution of nitrogen gas is usually observed by its hydrolysis in boiling water. It is also sensitive to impact and shock. If density of the compact of tetrazene is more, then, it loses its detonation property. It means that its detonation property is observed only when it is not compacted. It indicates that it requires more oxygen and hence, when it is compacted, it will have less oxygen exposure for burning to detonation.

Uses:

Even though it is not suitable for filling detonators, tetrazene is used in ignition caps.

Secondary Explosives:

Secondary explosives are also high explosives but they do not undergo detonation readily by heat or shock. However, on initiation, the secondary explosives rapidly and instantly decompose into stable components. The decomposition of secondary explosive, $C_3H_6N_6O_6$ (RDX) into stable components is illustrated below in an equation.

$$C_3H_6N_6O_6 \longrightarrow 3CO + 3H_2O + 3N_2$$

During the decomposition of RDX, considerable amounts of gases evolve along with heat. RDX can be a violent explosive if stimulated with a primary explosive. Sometimes, secondary explosives are stable and they can't undergo detonation; rather, they can be set on fire.

Some examples of secondary explosives are TNT, tetryl, picric acid, nitrocellulose, nitroglycerine, nitroguanidine, HMX, and TATB.
The chemistry of secondary explosives is outlined along with their structures below one by one.

Nitroglycerine:

Synthesis:

The first preparation of nitroglycerine ($C_3H_5N_3O_9$) was reported by Italy's Ascanio Sobrero in 1846. Thus, its preparation involves adding glycerol to a mixture of sulfuric and nitric acids.

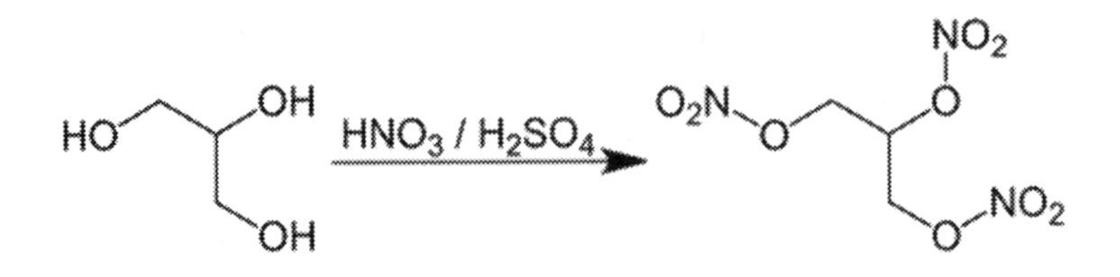

Properties:

Nitroglycerine is toxic to handle and produces toxic products on detonation. It also causes headaches. It is insoluble in water but soluble in most of the organic solvents. It also forms gels readily with nitrocellulose.

Uses:

Nitroglycerine is a powerful secondary explosive. It has high shattering effect and the high shattering effect is called brisance. It is commercially used. It also provides enough and high energy in propellant compositions. It is used as solid rocket propellants in combination with nitrocellulose and stabilizers.

Nitrocellulose:

Synthesis:

The discovery of nitrocellulose was reported by C.F. Schonbeim at Basel and R. Bottger at Frankfurt-am-main during 1845-47. Abel improved the stability of nitrocellulose in 1865. Nitrocellulose materials are prepared from cotton. Thus, cellulose is nitrated in acidic conditions to prepare nitrocellulose as shown below.

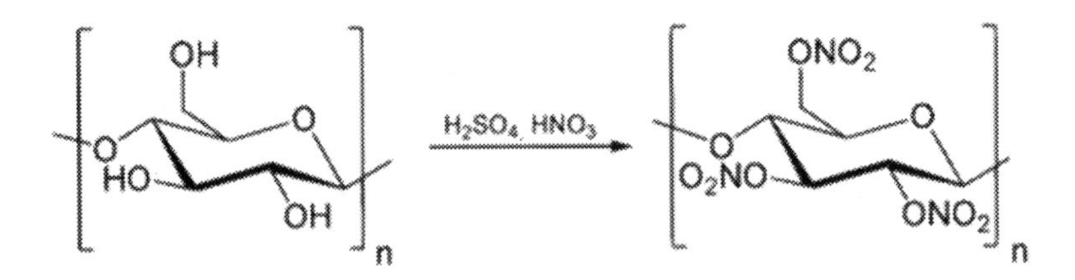

Properties:

It is a fluffy white solid when its synthesis is from cotton. It does not undergo melting; rather, it directly decomposes in the region of 180°C it is very sensitive to electrostatic discharge. It is soluble in organic solvents to form a gel.

Uses:

Since it is a polymeric in nature, it has good physical properties. It is used in gun propellants and double-base rocket propellants.

Picric Acid:

Synthesis:

Picric acid is prepared by nitration of phenol as represented by the chemical equation below.

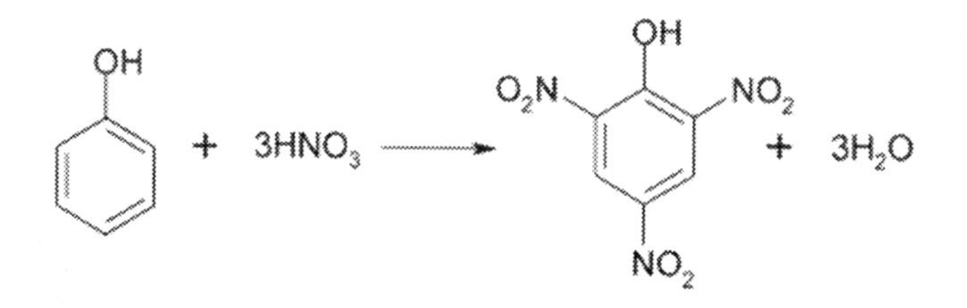

Properties:

It is a strong acid. It is soluble in hot water, alcohol, ether, benzene, and acetone. It attacks common metals; however, aluminum and tin metals are stable against picric acid. It is also toxic.

Tetryl:

Synthesis:

Its preparation involves two steps. The first step is treating the dinitrochlorobenzene with methylamine in alcohol solvent and the second step is to treat with Nitric acid in acidic conditions.

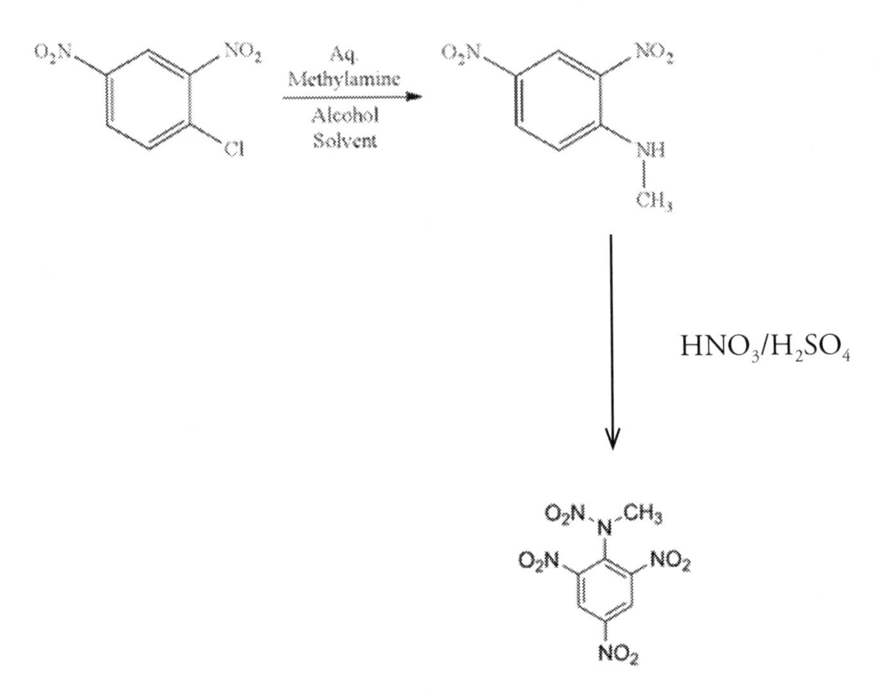

Properties:

It is a pale yellow crystalline solid. Because of its crystalline nature, it has a melting temperature of 129°C. It is used in pressed pellet forms as primers for explosive compositions. It is toxic to handle.

TNT:

Synthesis:

Its first preparation was reported by Wilbrand in 1863. Its preparation involves nitration of toluene using fuming nitric acid in the presence of sulfuric acid as represented below in a chemical reaction.

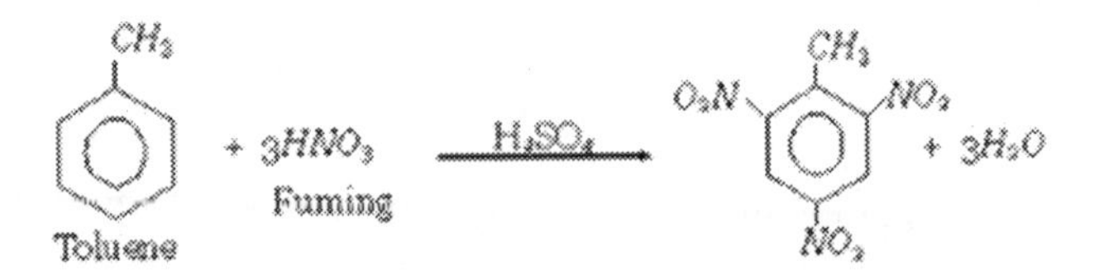

Properties:

It is insoluble in water but slightly soluble in alcohol. It is completely soluble in benzene, toluene, and acetone. It is unstable in alkalis and amines.

Nitroguanidine or Picrite:

Synthesis:

Its first preparation was reported by Jousselin in 1877 and its preparation involves dissolving dry guanidine nitrate in fuming nitric acid and passing nitrous oxide through the solution. Finally, the solution is poured into water to precipitate nitroguanidine. Its structure is shown below.

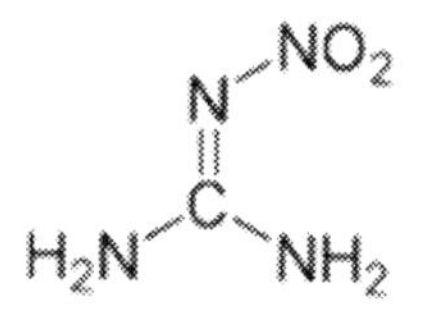

Properties:

Below the melting point it is a stable solid. But, upon melting it produces ammonia, water vapor, and solid product. It is soluble in hot water and alkalis. But it is insoluble in ethers and cold water.

PETN:

Synthesis:

Reaction of pentaerythritol with nitric acid yields PETN (PentaErythritol Tetranitrate, $C_3H_8N_4O_{12}$) as represented by the following equation.

$$C(CH_2OH)_4 + 4\ HNO_3 \rightarrow C(CH_2ONO_2)_4 + 4\ H_2O$$

Properties:

It is most stable and less reactive explosive among nitric esters. Even when stored at 100°C for a very long time, it does not show any trace decomposition. It has great shattering effect.

RDX:

Synthesis:

Its first preparation was known in 1899 by Henning for medicinal use and it was later used as an explosive in 1920 by Herz. Its synthesis involves a reaction between hexamethylenetetramine and nitric acid, ammonium nitrate and acetic anhydride as represented by a chemical reaction.

It is also known as Hexogen, Cyclonite, and Cyclotrimethylenetrinitramine.

$$(CH_2)_6N_4 + 4HNO_3 + 2NH_4NO_3 + 6(CH_3CO)_2O$$

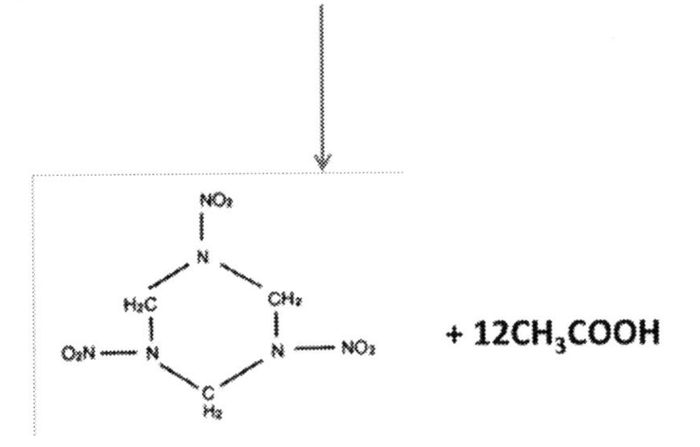

Properties:

It is a white crystalline solid and has a melting point of 204°C. It was used by military persons during World War II due to being more chemically and thermally stable. It has low solubility in organic solvents but recrystallization is usually achieved from acetone.

HMX:

Its different names are Octogen and cyclotetramethylenetetranitramine. Its structure is shown below.

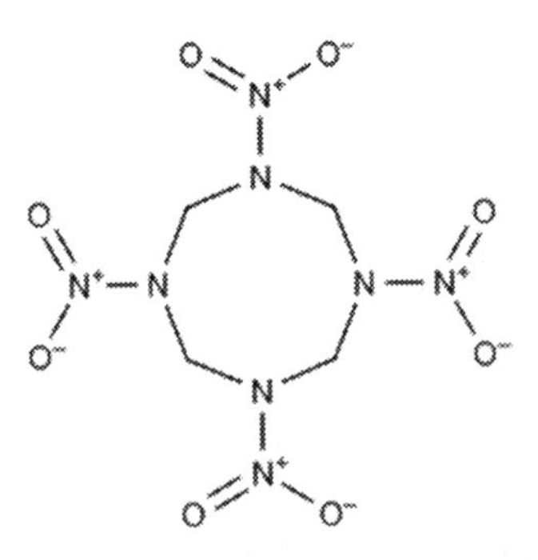

It is a white and crystalline solid and it appears to have four different crystalline forms differing from one another. Among them, the beta-form is employed in secondary explosives due to being least sensitive to impact. It is non-hygroscopic and insoluble in water.

TATB:

It has other name and it is 1,3,5-triamino-2,4,6-trinitrobenzone. Its structure is shown below.

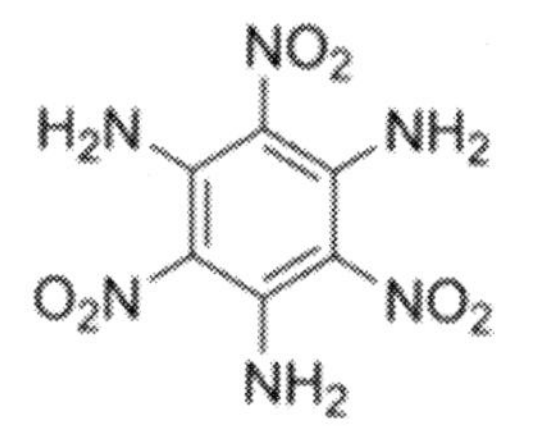

Its color is yellow-brown. Its decomposition temperature is well below its melting point and its decomposition is rapid. It is known as a heat-resistant explosive due to its excellent thermal stability in the range 260 – 290°C.

Its crystal structure gains much importance due to its many unusual features. Molecules are arranged in planner sheets in the unit cell. Strong intra and intermolecular hydrogen bonding exist between these planner sheets. Its structure resembles a graphite-like lattice structure and it has lubricating and elastic properties.

HNS:

Hexanitrostibene is abbreviated as HNS. It has the following unique characteristics.

a. It is a heat-resistant explosive
b. It is also resistant to radiation
c. It is insensitive to an electric spark
d. it is less sensitive to impact.

Its structure is shown below.

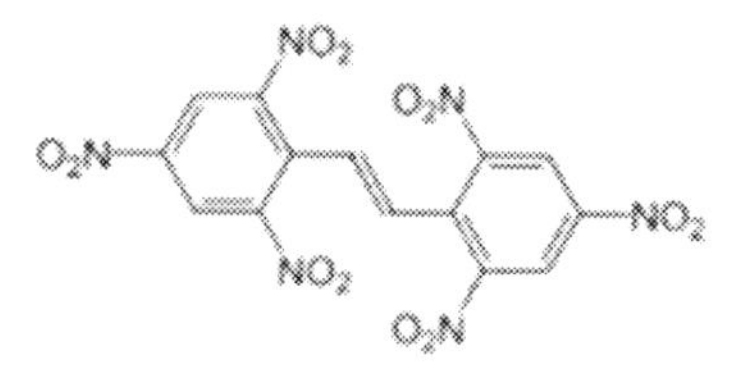

NTO:

Its attractive characteristics are:

a. It is a new energetic material
b. It has a high heat of reaction
c. It is more stable than RDX.
d. It is a substitute for ammonium nitrate or ammonium perchlorate in solid rocket propellant.
e. Its decomposition products do not contain acid such as HCl.

Its structure is shown below.

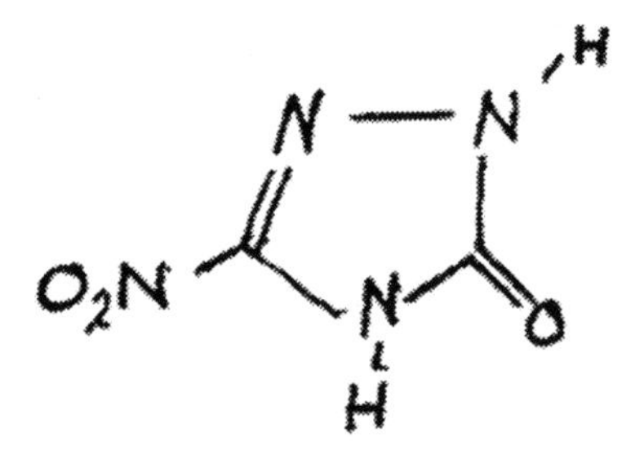

TNAZ:

Its other name is 1,3,3-trinitroazetidine. Its first preparation was known in 1983 in Fluorochem Inc. It is a white crystalline solid. It is soluble in most common organic solvents such as acetone, methanol, ethanol, tetra chloromethane, and cyclohexane. Its structure is shown below.

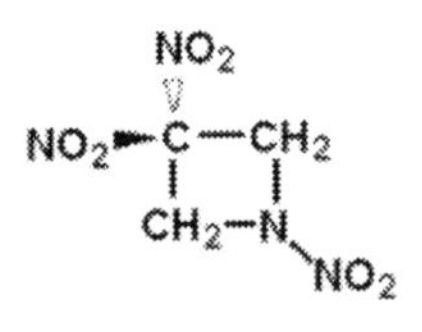

Propellants:

Propellants are combustible materials thus they need oxygen for their combustion. They do not explode; rather, they only burn. Some examples of propellants are black powder, smokeless propellant, blasting explosives, and ammonium nitrate.

Thermochemistry of Explosives:

On ignition explosives undergo rapid decomposition. Therefore, their thermochemistry play important role to understand

a. type of chemical reaction
b. energy changes and
c. mechanisms and kinetics of decomposition reaction.

Explosives on decomposition yield smaller molecules such as N_2, H_2O, CO, and CO_2, which are of course stable in nature.

Oxygen balance:

All the explosives are fuels in nature and hence, they have or require amount of oxygen for complete oxidation. For example, Trinitrotoluene (TNT) is an explosive with some oxygens are present in it, as shown its structure below.

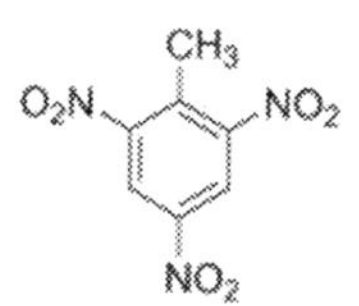

Thus, its molecular formula is $C_7H_5N_3O_6$. On detonation TNT is fully oxidized to form the gases carbon dioxide, water and nitrogen as shown in the reaction below.

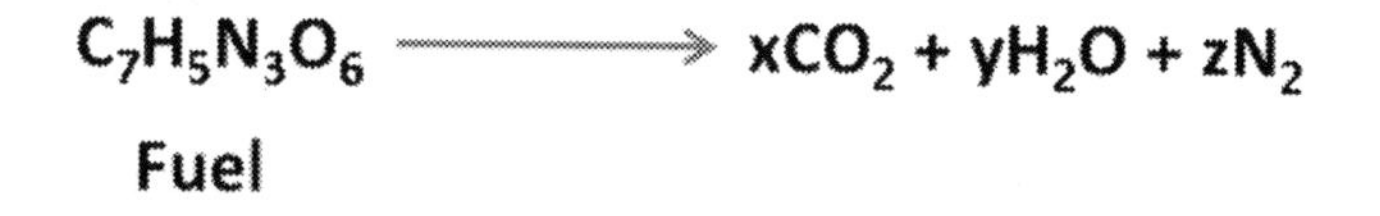

To balance the above equation Oxygen consumed from the air to be included to balance the above equation. Therefore, we write the equation as

$$\underset{\text{Fuel}}{C_7H_5N_3O_6} \xrightarrow{O_2} xCO_2 + yH_2O + zN_2$$

Accordingly, it is assumed that for complete combustion, oxidizer valence (O) to fuel valence (F) should be equal to one. i.e.

$$\frac{\textbf{Oxidizing Valence}}{\textbf{Fuel Valence}} = 1$$

We can calculate fuel valence from fuel, in this case it is TNT. Substituting this value in the above equation, we will get oxidizing valence and from this oxidizing valence, we can find out moles of oxygen required.

Estimation of Fuel Valence:

Valences for Carbon, Hydrogen, Oxygen and Nitrogen are assumed as +4, +1, -2 and zero due to their conversion into CO_2, H_2O and N_2 molecules on detonation, in which the valences of carbon, oxygen, hydrogen and nitrogen are +4, -2, +1 and zero. Therefore, fuel valence calculation in the case of TNT is as follows.

7C	= 7 x (+4)	= +28
5H	= 5 x (+1)	= +5
3N	= 3 x (0)	= 0
6O	= 6 x (-2)	= -12
$C_7H_5N_3O_6$		= +21

Therefore, fuel valence of TNT molecule is +21 and substituting this value in the above formula, we will get

$$\frac{\textbf{Oxidizing valence}}{\textbf{1(+21)}} = \textbf{1}$$

Therefore, oxidizing valence of O_2 molecule = +21
Substituting the valence of Oxygen as -2, we will get
Valence of oxygen = -4

Therefore,

$$\frac{5.25\ (-4)}{1(+21)} = 1$$

The above equation indicates that for every mole of TNT 5.25 moles of O_2 molecule is required. Therefore, for complete oxidation of TNT, the redox reaction can be written as follows.

$$\underset{\text{Fuel}}{C_7H_5N_3O_6} \xrightarrow{5.25\ O_2} 7CO_2 + 2.5H_2O + 1.5N_2$$

In the simplified version of oxygen balance calculation for TNT, $C_7H_5N_3O_6$ is given as

$$\text{Oxygen Balance} = \frac{[6-(7x2)-(5/2)]X1600}{\text{Molecular weight of TNT}}$$

$$\text{Oxygen Balance} = \frac{-16{,}800}{227} = -74\%$$

The negative sign of the result indicates that for the completion of detonation of TNT 74% oxygen is required in the reactants side.

In general, for formula of $C_aH_bN_cO_d$, the oxygen balance becomes

$$\text{Oxygen balance} = \frac{[d-(2a)-(b/2)]x\ 1600}{\text{Molecular weight of } C_aH_bN_cO_d}$$

Example 1:

Calculation of oxygen balance of NH_4NO_3:
As we know oxygen balance for the formula $C_aH_bN_cO_d$

$$\text{Oxygen balance} = \frac{[d-(2a)-(b/2)] \times 1600}{\text{Molecular weight of } C_aH_bN_cO_d}$$

Therefore, oxygen balance for NH_4NO_3 becomes

$$\text{Oxygen balance} = \frac{[3 - (2x0)-(4/2)] \times 1600}{80.04} = \frac{1600}{80.04} = +19.99\%$$

The positive sign indicates that during decomposition of NH_4NO_3, it releases 19.99% of oxygen.

Table 2 below summarizes oxygen balance in % weight for various explosives.

Table 2: Oxygen balance of various explosives

Explosives	Empirical formula	Molecular weight	Oxygen balance, % weight
Nitroglycerine	$C_3H_5N_3O_9$	227.0865	+3.50
RDX	$C_3H_6N_6O_6$	222.1163	-21.60
HMX	$C_4H_8N_8O_8$	296.155	-21.62
Nitroguanidine	$CH_4N_4O_2$	104.068	-30.70
Picric acid	$C_6H_3N_3O_7$	229.1039	-45.40
Tetryl	$C_7H_5N_5O_8$	287.1433	-47.39
TATB	$C_6H_6N_6O_6$	258.148	-55.80

Decomposition Products of Explosives:

Decomposition products of explosives can be Carbon Monoxide and/or Carbon dioxide and/or carbon from the carbon constituent, hydrogen and/or water from the hydrogen constituent and nitrogen molecule from the nitrogen constituent. Therefore, in order to predict the decomposition products of explosives there are rules known and these rules are very briefly outlined with an example at least in the following section.

Kistiakowsky – Wilson Rules:

According to the Kisitakowsky – Wilson Rule (K – W rule),

a. Carbon atoms are converted into Carbon monoxide
b. After the conversion of carbon atoms into carbon monoxide, remaining oxygen (if at all) oxidizes hydrogen into water.
c. Still oxygen remains after the reactions of "a" and "b", carbon monoxide is converted into carbon dioxide.
d. All the nitrogen is converted into nitrogen molecule, N2 gas.

Application of Kistiakowsky – Wilson Rules (K – W rules):

What are the decomposition products of Nitroglycerine ($C_3H_5N_3O_9$)? Applying Kistiakowsky – Wilson rules,

a. 3 carbon atoms are converted into 3CO
b. 5 hydrogen atoms are oxidized into $2.5H_2O$
c. Remaining 3.5 oxygen atoms oxidize 3 CO into 3 CO_2
d. 3 nitrogen atoms are converted into 1.5 N_2 molecule (gas).

Therefore, according to Kistiakowsky – Wilson Rule, the decomposition products of Nitroglycerine are $3CO_2 + 2.5H_2O + 1.5N_2 + 0.5O$.

Table 3 below summarizes the decomposition products of some explosives using Kistiakowsky – Wilson Rules:

Table 3: Kistiakowsky-Wilson Rules decomposition products of explosives

Explosive	Decomposition Products (K – W rules)
Nitroguanidine	$CH_4N_4O_2 \rightarrow CO + H_2O + H_2 + 2N_2$
HMX	$C_4H_8N_8O_8 \rightarrow 4CO + 4H_2O + 4N_2$
RDX	$C_3H_6N_6O_6 \rightarrow 3CO + 3H_2O + 3N_2$
Nitroglycerine	$C_3H_5N_5O_9 \rightarrow 3CO_2 + 2.5H_2O + 1.5N_2 + 0.5O$

Modified Kistiakowsky – Wilson Rules:

If oxygen balance is lower than -40, then Kistiakowsky – Wilson Rules can't be used. Under this circumstance, modified Kistiakowsky – Wilson Rules are applicable.

According to modified Kistiakowsky – Wilson Rules,

a. Hydrogen atoms are converted into water first.
b. If oxygen atoms remain after the first step, "a", carbon atoms are converted into Carbon Monoxide.
c. After the first two steps, still oxygen atom(s) is available, Carbon monoxide is converted into Carbon dioxide
d. All the nitrogen atoms are converted into Nitrogen molecule (N_2 gas).

Application of Modified Kistiakowsky – Wilson Rules:

What are the decomposition products of picric acid, $C_6H_3N_3O_7$ (applying modified Kistiakowsky – Wilson Rule)?

Applying modified Kistiakowsky – Wilson Rules,

a. 3 hydrogen atoms are converted into 1.5 H_2O
b. From the remaining 5.5 oxygen atoms, 5.5 carbon atoms consume to become 5.5CO.
c. 0.5 carbon atom remains as 0.5 carbon.
d. 3 nitrogen atoms are converted into 1.5 N_2 molecule (gas).

Thus,
Based upon modified Kistiakowsky – Wilson Rules, the decomposition products of picric acid are shown below in an equation.

$$C_6H_3N_3O_7 \rightarrow 1.5H_2O + 5.5CO + 0.5C + 1.5N_2$$

Table 4 below summarizes the decomposition products of some explosives (applying the modified Kistiakowsky – Wilson Rules).

Table 4: Modified Kistiakowsky-Wilson Rules decomposition products of explosives

Explosive	Decomposition Products (modified K – W rules)
TNT	$C_7H_5N_3O_6 \rightarrow 2.5H_2O + 3.5CO + 3.5C + 1.5N_2$
Tetryl	$C_7H_5N_5O_8 \rightarrow 2.5H_2O + 5.5CO + 1.5C + 2.5N_2$
Picric Acid	$C_6H_5N_3O_7 \rightarrow 1.5H_2O + 5.5CO + 0.5C + 1.5N_2$

Springall Roberts Rules:

Springall Roberts rules consider unmodified Kistiakowsky – Wilson rules and two additional conditions (e and f) are included. Thus, Springall Robert Rules are

a. Carbon atoms are converted into Carbon monoxide
b. After the first step, if oxygen atom(s) remains, then hydrogen atom is oxidized into water.
c. After the above two steps, if oxygen atom remains, then Carbon monoxide is converted into Carbon dioxide.
d. All the nitrogen atoms are converted into Nitrogen molecule (gas).
e. One out of three CO formed is converted into Carbon and carbon dioxide
f. One out of six CO formed in converted into Carbon and H_2O.

Application of Springall Robert Rules:

What are decomposition products of Picric acid? (applying Springall Robert Rules).

Applying the Springall Robert Rules, the decomposition products of Picric acid, $C_6H_3N_3O_7$ are

a. 6 carbon atoms are converted into 6CO
b. Remaining one oxygen atom oxidizes Hydrogen into H_2O
c. No more oxygen atom remains and hence, there is no possibility for conversion of carbon monoxide into Carbon dioxide.
d. 3 Nitrogen atoms are converted into $1.5N_2$ molecule (gas).
e. 2CO are converted into $C + CO_2$
f. CO is converted into C and $0.5H_2O$.

Thus, the decomposition products of picric acid applying Springall Robert Rules are

$$C_6H_3N_3O_7 \rightarrow 3CO + 2C + CO_2 + 1.5H_2O + 1.5N_2$$

So far, the decomposition products of explosives based upon two major rules are studied. However, the two major rules did not tell about the heat of reaction. Therefore, it is essential to calculate the heat of explosive reactions. In the following section, it will be the subject of calculation of the heat of explosive reactions.

Heat of Reactions (explosive versus combustion reaction):

Exothermic Reaction:

When heat is released during the reaction is called exothermic reaction. Heat of Exothermic reaction is negative value

Endothermic Reaction:

When heat is consumed during the reaction is called endothermic reaction. Heat of endothermic reaction is positive value.

Heat of Reaction of Explosive Materials:

Heat of reaction containing explosive materials is total heat evolved during the oxidation of explosive materials in excess amount of oxygen. Thus, the main products of reaction of explosive materials in the excess amount of oxygen are

a. Conversion of C into CO_2
b. Conversion of H into H_2O
c. Conversion of N into N_2
d. Conversion of Sulfur into SO_2

In other hand, the main products of reaction of explosive materials in the oxygen lean environment are

a. Conversion of C into CO and/or C
b. Conversion of H into H_2
c. Conversion of N into N_2

Also, heat of reaction of explosive materials in the excess amount of oxygen is called heat of combustion and heat of reaction of explosive materials in the oxygen lean environment is called heat of explosion. Thus, heat released during heat of combustion is higher than that of heat released during heat of explosion. In both cases, energy released during bond formation of new molecules is usually greater than that of energy required during bond breaking of explosive chemicals.

Standard internal energy of formation of products and reactants are used when the heat of reaction of explosive materials is carried out at constant volume where as standard enthalpy of formation of products and reactants are used when the heat of reaction of explosive materials is carried out at constant pressure.

Calculation of Heat of Reaction (heat of combustion in excess O_2 and heat of explosion in O_2 lean):

Example 1: Estimate heat of explosion of RDX, $C_3H_6N_6O_6$ and given are

Standard heat of formation of RDX = +62 KJ/mol
Standard heat of formation of CO = -110 KJ/mol
Standard heat of formation of H_2O = -242 KJ/mol

Note: Standard heat of formation of N_2, O_2, H_2 and all other elements are zero.

Solution:

The stoichiometric reaction of explosion of RDX is

$$C_3H_6N_6O_6 \rightarrow 3CO + 3H_2O + 3N_2$$

Heat of Detonation of RDX = {Sum of heat of formation of Products} - {Sum of heat of formation of reactant}

Sum of heat of formation of products

= 3 x (-110) + 3 x (-242) = -330 – 726 = -1056 KJ/mol

Sum of heat of formation of reactant

= 1 x 62 = 62 KJ/mol

Therefore, heat of Detonation = -1056 – (+620) = -1118 KJ/mol

$$\text{Heat of explosion} = \frac{\text{Heat of Detonation x 1000}}{\text{Molecular weight of RDX}} = \frac{-1118 \times 1000}{222}$$

= -5036 KJ/Kg

Example 2:

Estimate the heat of combustion of RDX and given are

Standard heat of formation of RDX = +62 KJ/mol
Standard heat of formation of CO_2 = -393.7 KJ/mol
Standard heat of formation of H_2O = -242 KJ/mol

Note: Standard heat of formation of N_2, O_2, H_2 and all other elements are zero.

Solution:

The stoichiometric reaction of combustion of RDX is

$$2C_3H_6N_6O_6 + 3\,O_2 \rightarrow 6CO_2 + 6H_2O + 6N_2$$

Heat of Detonation of RDX = {Sum of heat of formation of Products}
- {Sum of heat of formation of reactant}

Sum of heat of formation of products

= 6 x (-393.7) + 6 x (-242) = -2362.2 – 1452 = -3814.2 KJ/mol

Sum of heat of formation of reactant

= 2 x 62 = 124 KJ/mol

Therefore, heat of Detonation = -3814.2 – (+124) = -3938.2 KJ/mol

$$\text{Heat of Combustion} = \frac{\text{Heat of Detonation x 1000}}{\text{Molecular weight of RDX}} = \frac{-3938.2 \times 1000}{222}$$

= -17,739.64 KJ/Kg

Note:
From Example 1 and Example 2, it can be evidenced and concluded that heat of combustion is greater than that of heat of explosion for a given explosive material.

Example 3:
Estimate the heat of explosion and combustion of Nitroglycerine, $C_3H_5N_3O_9$ and given are

Standard heat of formation of Nitroglycerine = -380 KJ/mol
Standard heat of formation of CO_2 = -393.7 KJ/mol
Standard heat of formation of H_2O = -242 KJ/mol

Note: Standard heat of formation of N_2, O_2, H_2 and all other elements are zero.

Solution:
The stoichiometric reaction of explosion of nitroglycerine is

$$C_3H_5N_3O_9 \rightarrow 3CO_2 + 2.5H_2O + 1.5N_2 + 0.5\ O$$

Heat of Detonation of Nitroglycerine = {Sum of heat of formation of Products} - {Sum of heat of formation of reactant}

Sum of heat of formation of products

= 3 x (-393.7) + 2.5 x (-242) = -1181.1 – 605 = -1786.1 KJ/mol

Sum of heat of formation of reactant

= 1 x -380 = -380 KJ/mol

Therefore, heat of Detonation = -1786.1 – (-380) = -1406.1 KJ/mol

$$\text{Heat of Explosion} = \frac{\text{Heat of Detonation x 1000}}{\text{Molecular weight of Nitroglycerine}} = \frac{-1401.1 \text{ x } 1000}{227}$$

= -6172.25 KJ/Kg

The stoichiometric reaction of combustion of Nitroglycerine is
The stoichiometric reaction of explosion of nitroglycerine is

$$C_3H_5N_3O_9 \rightarrow 3CO_2 + 2.5H_2O + 1.5N_2 + 0.5\ O$$

Heat of Combustion of Nitroglycerine = {Sum of heat of formation of Products} - {Sum of heat of formation of reactant}

Sum of heat of formation of products

= 3 x (-393.7) + 2.5 x (-242) = -1181.1 – 605 = -1786.1 KJ/mol

Sum of heat of formation of reactant

= 1 x -380 = -380 KJ/mol

Therefore, heat of Detonation = -1786.1 – (-380) = -1406.1 KJ/mol

$$\text{Heat of Combustion} = \frac{\text{Heat of Detonation x 1000}}{\text{Molecular weight of Nitroglycerine}} = \frac{-1401.1 \text{ x } 1000}{227}$$

$= -6172.25$ KJ/Kg

Note:

From Example 3, it can be concluded that heat of explosion is same as heat of combustion for nitroglycerine explosive substance.

Explosive Power and Power Index:

During the explosive reaction, both heat and gases are liberated. Therefore, independently, heat of explosion (Q) and volume of gas (V) can be calculated. Thus, Explosive power is given as

Explosive Power = Heat of Explosion (Q) x Volume of Gas (V)

Power index is defined as the ration of explosive power to standard explosive (picric acid) as shown below

$$\text{Power Index} = \frac{\text{Heat of Explosion (Q) x Volume of Gas (V)}}{\text{Q of Picric acid x V of Picric acid}} \text{ x } 100$$

Questions:

1. State 3 types of classification of explosion with an example for each type.
2. Explain 3 types of chemical explosives with examples.
3. Define primary explosives and provide two examples with your explanation.
4. What are secondary explosives? Provide a couple of examples with your explanation.
5. What is TNT? What is the chemistry (preparation, properties and uses) of TNT.
6. What is combustion reaction? Give an example.
7. State thermochemistry of explosives with an example.
8. State Kistiakowsky – Wilson rule and its modified version

9. What is the application of Kistiakowsky – Wilson rules?
10. What is the application of Springall Robert rules?
11. Define explosive power.
12. Define power index.

Metallurgy

Objectives:

1. To define Metallurgy
2. To define three major extractive metallurgy
3. To describe pyrometallurgical process
4. To outline two methods of reduction processes
5. To describe need for Pyrometallurgical refining
6. To elaborate hydrometallurgical process
7. To elaborate various purification methods for metals
8. To elaborate electrometallurgical process
9. To illustrate extraction of important metals

Metallurgy is the science and technology that leads to extract metals and alloys from their minerals. Thus, extractive metallurgy technique has been known to extract metals and alloys from their minerals.

There are three major processes known for extractive metallurgy and these are

Pyrometallurgical process
Hydrometallurgical Process and
Electrometallurgical Process.

Pyrometallurgical Process:

Pyrometallurgical process involves the following steps.

1. Drying
2. Calcination
3. Sintering
4. Pyrometallurgical Reduction
    - Reducing agents
    - Reduction of non-oxide compounds
5. Pyrometallurgical Refining

1. Drying:

20 – 30 weight percent of ores consist of bulk water and volatile compounds. Therefore, drying process is used first to remove the bulk water and volatile compounds to concentrate the ores by 20 to 30 volume percent. Either heating in air or vacuum heating is explored to evaporate water and removal of organic volatile compounds respectively. In this process, weight loss of ore is observed and the percentage of weight loss is known as Loss On Ignition (LOI). As evaporation of water is endothermic (heat of evaporation of water is 44 KJ/mol) sufficient time is allowed at the drying temperature for the complete removal of water molecules.

2. Calcination:

It involves high temperature operation, which is usually higher than that of drying temperature. Thus, calcination step removes chemically bound water or decomposition of carbonates into CO_2. This step is important for carbonate minerals and one typical example includes lime from limestone. At 1 atm pressure, decomposition of limestone into lime is usually at 1000°C as shown the chemical equation below.

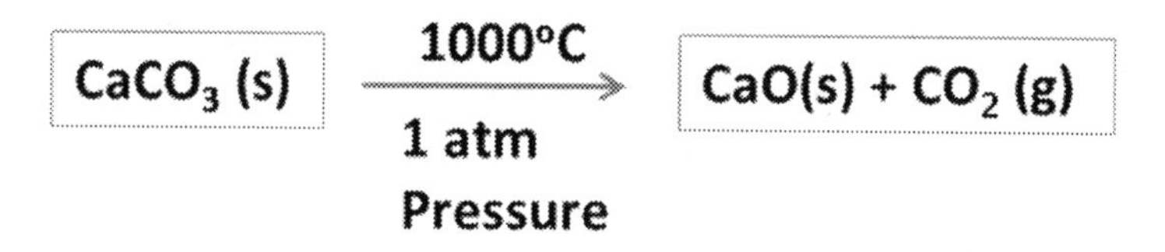

3. Sintering:

Sintering is the process by which partial reduction of material and agglomeration (or coarsening) of relatively fine particles take place. Therefore, sintering requires ever higher temperature than that of calcination step. In the case of iron ore, drying, calcination and sintering are simultaneously achieved. But, sintering of non-ferrous sulfide ores of lead, copper, zinc and other metals involve coarsening in addition to controlling the particle size and thus, dual function of particle size control and coarsening is achieved.

4. Pyrometallurgical Reduction:

After the first, second and third steps of drying, calcination and sintering respectively, the ores are concentrated into high concentration of oxide or sulfide. Therefore, at this stage, reduction of metal oxide/ metal sulfide into required metal is achieved using a reducing agent as shown by a general equation.

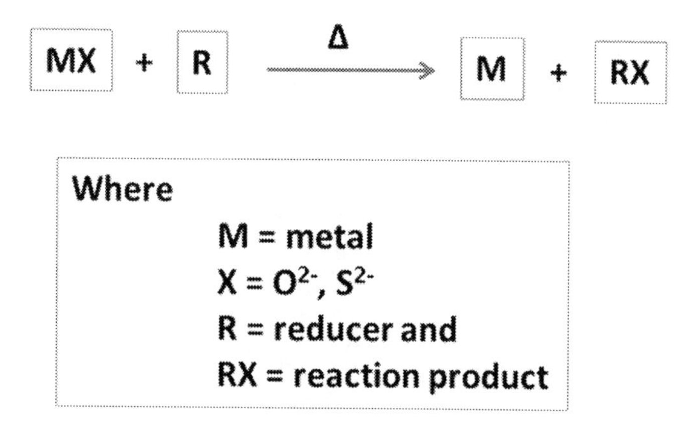

Hydrothermal reduction Process and Carbothermal reduction process:

In the pyrometallurgical reduction process, either hydrogen or carbon monoxide is explored as reducer in most cases. If hydrogen gas (H_2) is used as a reducer in the pyrometallurgical reduction process, this process is called hydrothermal reduction process. If carbon monoxide (CO) is used as a reducer in the pyrometallurgical reduction process, the process is called carbothermal reduction process. The chemical reactions that represent hydrothermal reduction process and carbothermal reduction process are shown below with formation of thermodynamically stable reaction products such as H_2O and CO_2 respectively.

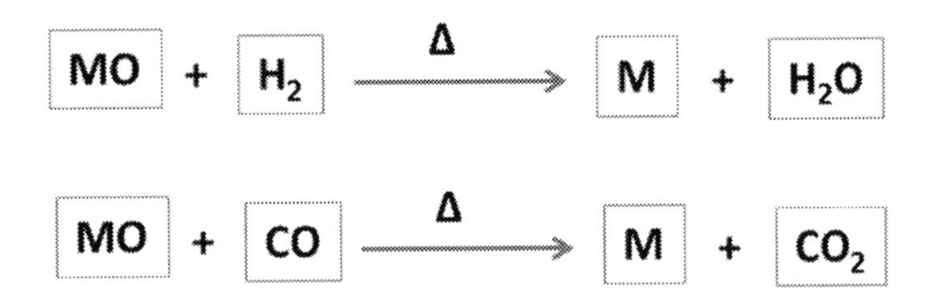

Metallothermal Reduction Process:

Like hydrothermal and carbothermal reduction processes, metallothermal reduction process is also explored in the pyrometallurgical reduction process. Thus, a metal can reduce other metal oxide into metal. The metallothermal reduction process can be understood by the following example. Magnesium metal is the reducer and magnesium metal is used to reduce either silica (SiO_2) or titania (TiO_2) into Silicon (Si) and Titanium (Ti) respectively by the formation of reaction product of MgO, which is thermodynamically stable product at the reducing temperature. The metallothermal reduction process is illustrated by the following chemical equation(s).

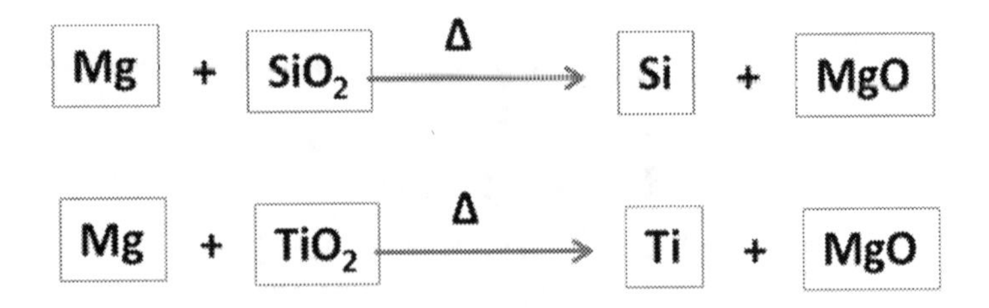

Yet another and known example for metallothermic reduction reaction is use of aluminum metal powder as a reducer to reduce iron oxide (Fe_2O_3) into molten iron and solid aluminum oxide powder at the operation temperature of metallothermic reduction reaction.

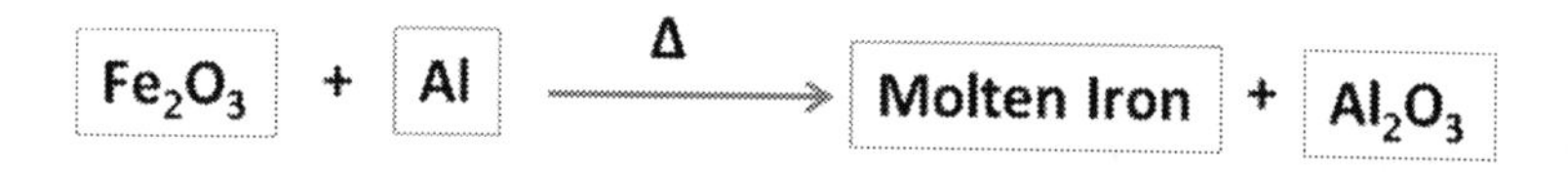

Metallothermic reaction is also called as Goldschmidt Reaction.

Metallothermal reduction of metal oxide has a serious problem associated with seizing of the process once the solid reaction product is covered at the surface. Therefore, gaseous reaction product is preferred way of manufacturing or reduction of metal oxides into metal.

Carbon as a reducer:

Carbon is a relatively cheap reducer and it has been explored to reduce metal oxides into metals in the pyrometallurgical process. Carbon can reduce refractory metal oxides into metal at elevated temperature. However, a major drawback of use of carbon powder as a reducer is that some metals tend to form stable metal carbides at the reducing temperature. In that case, hydrogen gas as a reducer is successfully used to reduce metal oxides into metal. But, the cost of hydrogen gas is higher than that of carbon powder.

Reduction of Non-oxide Compounds:

There are refractory metal oxides that are known to have high affinity to oxygen and hence, even by strong reducing agent can't reduce the metal oxides into metals. Examples for such oxides are UO_2, TiO_2, ZrO_2 and so on. In this case of difficulty in reducing metal oxides into the corresponding metals, conversion of the metal oxides into their halides in general and chlorides in particular followed by reduction of metal halides into metals. Conversion of metal oxides into the chlorides by chlorination or carbo-chlorination is usually achieved at higher temperature, and examples for direct chlorination and carbo-chlorination are shown below.

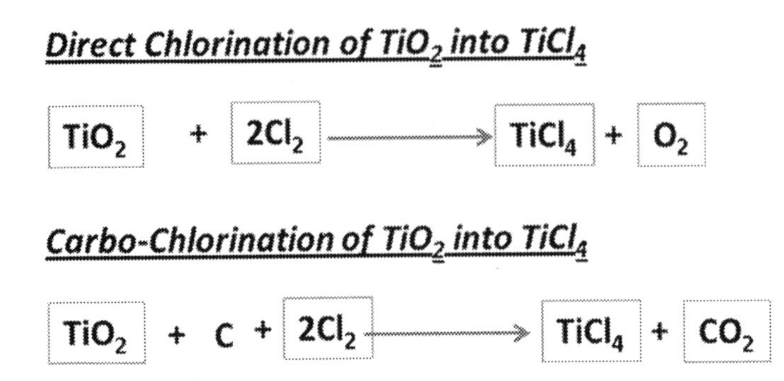

It is easier to reduce $TiCl_4$ into titanium by thermal reduction process.

In the case of lead metal (Pb) production, two metal compounds are reacted to get lead. Two metal compounds of lead are lead oxide (PbO) and lead sulfide (PbS). The formation of lead by reaction between lead oxide (PbO) and lead sulfide (PbS) is shown below.

Lead metal Production by two metal compounds route

5. Pyrometallurgical Refining:

Sometimes, reducing agent(s) used in the pyrometallurgical process can reduce inner wall of reaction vessel and hence, it can contaminate the metal produced during pyrometallurgical reduction process. Therefore, it is essential to further maximize the purity of metal that is produced by pyrometallurgical reduction process. This referred as pyrometallurgical refining. Liquation process and refining with gaseous reagents are known to improve the purity of metal(s).

Hydrometallurgical Processes:

Hydrometallurgical processes do need pre-treatments such as drying, calcination and roasting as discussed in the hydrometallurgical processes. Hydrometallurgical processes are classified as

1. Leaching process
2. Purification &
3. Metal deposition using aqueous solution.

Because of involvement of aqueous medium hydrometallurgical processes need room temperature or boiling temperature of water as their temperature of operation. Boiling temperature of water is to accelerate the kinetics of the processes.

Leaching Processes:

Leaching is essentially involving separation process using aqueous medium. Leaching process is a kinetically-controlled process. In order to achieve leaching process successfully, concentration of active metal component in the bulk should be higher than that of metal component at the solid mineral and water surface.

Leaching of a compound in aqueous environment is due to simple dissolution reaction in water or acid and alkali as shown below with examples on each one.

Simple aqueous dissolution of metal:

Hydration of $CuSO_4$ in pure water is an example for simple aqueous dissolution of copper metal ion.

Acid dissolution of Metal:

Acid dissolution of zinc oxide is an example for acid assisted leaching of zinc oxide, as illustrated in an equation below.

Alkali dissolution of metal:

Alkali dissolution of Alumina, Al_2O_3 is an example for alkali assisted leaching of alumina, as illustrated with a chemical equation below.

$$Al_2O_3(s) + 2OH^-(aq) \longrightarrow AlO_2^-(aq) + H_2O$$

Other reactions involving in the leaching of metal ions from the ores are anionic base exchange, oxidation-reduction method and water soluble complex formation reaction. The leaching of metal ions by these reactions is outlined very briefly below.

Leaching by anionic base exchange (metathesis) reaction:

Anionic with basic characteristics such as carbonate ion, CO_3^{2-} can exchange with tungstate anion, WO_4^{2-} as shown below.

$$CaWO_4(s) + CO_3^{2-}(aq) \longrightarrow WO_4^{2-}(aq) + CaCO_3(s)$$

Thus, solid metal tungstate is leached into solution by simple exchange or metathesis reaction.

Leaching by Oxidation-reduction method:

This method finds application to leach copper as cation from solid CuS ore by oxidation – reduction mechanism using aqueous ferrous salt. Thus, ferrous salt, Fe^{2+} reduces sulfide of copper into sulfur while Fe^{2+} gets oxidized into Fe^{3+} as shown below.

Leaching due to water soluble complex formation:

Again this method is useful to recover copper ion from solid CuO by leaching using NH^{4+} ion from aqueous solution as shown below.

Similarly, some chelating extractants used for recovery of copper by this mechanism are shown below.

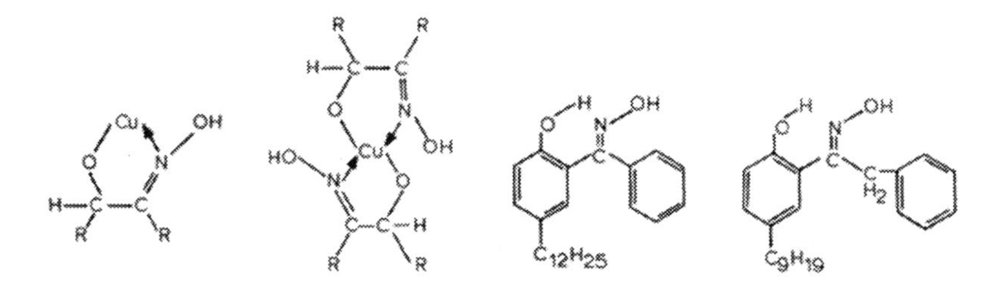

2. Solution Purification:

The leaching of desired metal ions into aqueous solution can take place with other impurity ions also. This is a problem associated with leaching process and hence, purification and recovery of particular and desired metal ions from the aqueous solution is, of course, needed even before deposition of metal. Only leaching of desired metal ion can be achieved from purified ore. Therefore, there are three major methods that are explored to purify the ores are

Precipitation method
Solvent extraction method
Ion-exchange method.

Purification by precipitation method:

The unwanted metal ions are usually separated from desired metal ion by precipitation of unwanted metal ions from the aqueous solution just after the leaching process. Thus, the concept of solubility product of precipitates plays a major role in the determination of complete separation of unwanted metal ions from the aqueous solution. Precipitation can be achieved by a few different methods. One of the common methods is the addition of appropriate cation or anion to precipitate specific compound(s) from the aqueous solution. Example includes the addition of sulfide to precipitate insoluble metal sulfide compounds. Another way is the evaporation of water so that at a certain high concentration of unwanted compound(s) can be precipitated. Also, yet another way is to alter the pH of solution to precipitate the unwanted compounds while the desired compound does not get precipitated. For example, iron, copper, cobalt and nickel are precipitated by the adjustment of pH of the solution to 2.5, 5.8, 8.3 and 9.4 respectively.

Purification by solvent extraction method:

The solvent extraction method is a chemical process for the purification of a given species from the aqueous solution. The immiscibility of organic solvent and, aqueous solution and stabilities of the metal species in the organic and water medium play prime roles to determine the purification by solvent extraction method. The chemistry of solvent of solvent extraction method involves solvation and exchange reaction. Solvation refers to transfer of neutral molecules between organic solvent and aqueous solution. The examples of organic solvents for solvation of neutral molecules are alcohols, ethers, esters, ketones and some phosphorus containing compounds. Exchange reaction refers to formation of chemical bond(s) between metal species and active compounds in the organic phase. The exchange reaction can be cationic or anionic. Thus, organic phase carries an extractant and a modifier. Thus, carrier for the extractant is about 90% of the solution and 90% of carrier acts as vehicle for carrying the active extractant. Examples for carrier vehicles are kerosene and naphthene.

Purification by ion-exchange method:

Ion-exchange method in the purification refers to extraction of metal ions between aqueous solution and a solid (insoluble resins). Thus, the chemistry involves in the ion-exchange method is very similar to that of ion-exchange method described in the solvent extraction method. The main difference is that the ion-exchange method involves ion-exchange reaction between solution and solid whereas the solvent extraction method involves ion-exchange reaction between two immiscible solutions. Thus, ion-exchange method here excludes the involvement of organic solvent. Therefore, the ion-exchange method involves adsorption of metal ions from aqueous solution by a solid resin and it is followed by elution. The ion-exchange method of purification is very similar to column chromatography. Two types of resins are explored in the ion-exchange method of purification of metals. One is cationic resin and another one is anionic resin. Mostly, cationic resins are either strong acids or weak acids with exchangeable protons with metals ion from the aqueous solution. Anionic resins are either mostly strong bases or weak bases. Strong bases are usually quaternary ammonium groups, weak bases are usually secondary or tertiary amine group bases. The use of ion-exchange resins (both caionic and anionic types) are usually selected by their capacity, selectivity with metal ions/anions present in the aqueous solution.

3. Metal Deposition:

After having gone through leaching and purification of metals, desired metal should be recovered from the concentrated solution. Thus, three useful reducing techniques have been explored for metal recovery. These are called cementation, hydrogen reduction and carbon adsorption.

Cementation:

Cementation, also called as metallic replacement is essentially useful to recover or purify noble or less reactive metals from reactive metals present in the aqueous solution. Cementation refers to electrochemical redox reaction between a metal and metal ions in the aqueous solution. A typical example for cementation is reduction of Cd^{2+} from aqueous solution by solid zinc metal as represented by an equation below.

The whole reduction – oxidation (redox) reaction takes place at microscopic levels in the aqueous solution when zinc metal is added to aqueous solution containing noble cadmium ions. The reduction of Cd^{2+} ions by zinc metal is determined by electromotive series of metals. Thus, more electropositive metals tend to reduce less electropositive metal ions in aqueous solution. Also, kinetics of the redox reaction is determined by difference in potentials of two metals that are involved in the redox reaction. Thus, more the difference is, kinetics of the redox reaction is more.

Another example for the recovery of copper metal from aqueous solution of copper salt is by iron metals as shown redox reaction between Cu^{2+} and Fe below.

Gaseous Reduction:

Recovery of metals from aqueous solution containing metal ions using reducing agents involve passing hydrogen gas or carbon monoxide or SO_2 gases through the aqueous solution. Among them, hydrogen is the most favorable reducing agent due to direct recovery of metal after reducing by hydrogen gas. Further treatment may be required if CO or SO_2 gas is used as reducing gas. The chemistry involved in the gaseous reducing of metal ions from aqueous solution is the replacement reaction. For example, reduction of Cu^{2+} (aq) ions by Hydrogen gas is shown below.

To accelerate the above chemical reaction towards right side, sometime high pressure hydrogen gas is employed.

Carbon Adsorption Method:

This method of recovery of metals is essentially explored for noble metals such as gold, silver, platinum. Thus, this method involves reduction of noble metal ions from aqueous solution by solid carbon at relatively low temperature. In a typical experiment of carbon adsorption method, it involves passing the aqueous solution after leaching step through carbon column such that noble metals such as gold/silver/platinum deposit on carbon particles by surface adsorption mechanism. Caustic solution is then passed through the metal saturated carbon column to recover metals from the solid carbon support.

Note:

Carbon in Pulp Process:

It is also possible to include solid carbon in the leaching process for recovery of noble metals such as Au, Ag, Pt and the combo-process is known as carbon in pulp process.

Electrometallurgical Process:

Electrometallurgical process also refers to electrolytic process. Thus, electrical energy is used to recover or forced to deposit metals from aqueous solution on cathode.

The electrochemical reaction involves oxidation of metal with loss of electrons and reduction of metal ions with gain of electrons in a galvanic cell as shown below in Fig.1.

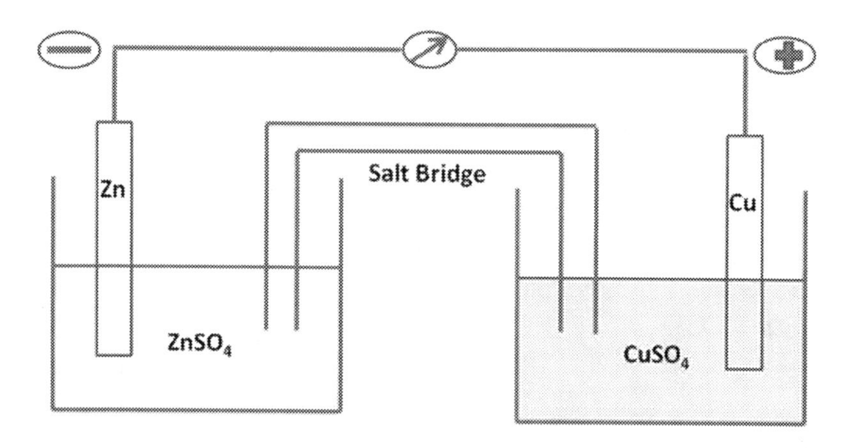

Fig.1: Electrochemical Galvanic Cell

Therefore, the valence state of metal is increased due to loss of electrons at the anode and the valence state of metal is decreased due the gain of electrons at the cathode and thus, cementation of metal ions takes place at the cathode.

The minimum voltage required from the saturated solution of ions for the deposition of metals is determined by electromotive series with respect to reference electrode. For example, copper electrode dipped in the copper sulfate solution serves as cathode whereas zinc metal dipped in zinc sulfate solution serves as anode as shown in the above figure. If the two metals from anode compartment and cathode compartment are connected, electrons flow from zinc anode to copper cathode is observed and thus, zinc metal dissolves to generate electrons and Zinc cations and copper metal accepts electrons to get Copper cations reduced and deposited at the cathode compartment.

Leaching followed by electrowinning process:

When acid leaching of oxide ore is carried out using sulfuric acid, sulfuric acid is consumed due to the following reaction.

Electrowinning is then followed to recover or deposit metal from metal sulfate aqueous solution at the cathode compartment of the galvanic cell. The reduction of metal sulfate that was obtained in the leaching process is shown below.

Interestingly, the SO_4^{2-} formed due the cathode reduction process of metal sulfate can be supplied to anode compartment where in sulfuric acid is regenerated by water oxidation process as shown below.

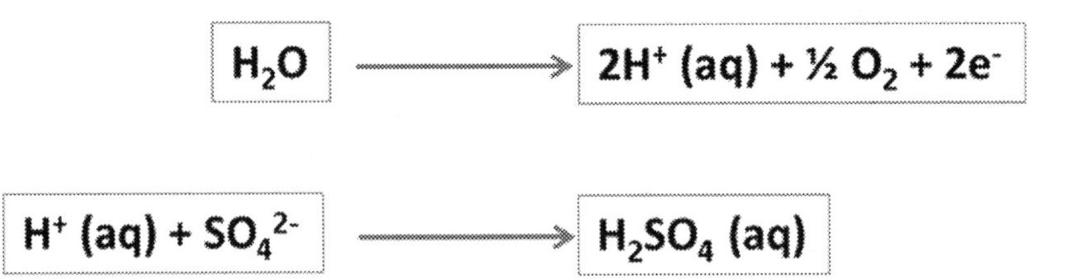

Thus, the acid leaching process and electrowinning process can be coupled together such a way ore is provided in the leaching process and metal is recovered in electrowinning process without supply of any other chemicals but of course, using galvanic cell in the latter process.

In the case of electrowinning process, the minimum voltage required to achieve the electrochemical process successfully is 1.25 to 1.75 V only. By the electrowinning process, deposition of metal at the cathode can be further improved by proper selection of cathode such that bonding between metal deposition and cathode should be loose enough.

For copper electrowinning process, the preferred cathodes are titanium, stainless steel and copper metal itself. Aluminum is used to recover zinc and titanium metals. For recovery of manganese and cobalt, stainless steel cathode is used.

Electro refining:

Electro refining is the purification process which is better purification process over other purification processes described earlier. Thus, concentrated metals can be recovered first and separately, impure metals can also be removed by the electro refining process of electrolysis method.

Molten Salt Electrolysis:

If metals having more electropositive in the electromotive series than manganese, then, the metals can't electro refined from aqueous solution due to electrolysis of water into hydrogen gas production rather than metal ions reduction at the cathode compartment. This should be overcome i.e. always electrolysis of water into hydrogen should be overcome in order to achieve reduction of metal ions from the aqueous solution. Examples of metals such as Aluminum, magnesium lithium, beryllium having more electropositive than water reduction into hydrogen gas production and therefore, these metal ions can't be reduced into metals from their aqueous solution in the cathode compartment. However, this problem can be avoided of water is eliminated. Thus, molten salt of these metals or ores of these metals can be successfully reduced into the metals at the cathode. Hence, this type of electrolysis of salts is called molten salt electrolysis of deposition or recovery of metals. This is illustrated with aluminum as an example below.

At the cathode compartment

In this case, other metal ions such as iron, silicon can also be reduced into metals along with aluminum. Therefore, it is highly desirable to go through the ores with Bayer process first to get as concentrated ores as possible to get highly pure aluminum metals by molten salt electrolysis.

Extraction of Some Important Metals:

The production and refining of naturally occurring ores depend upon the nature of chemistry of metals present in the ores. In this section, thus, extraction of gold, refining of iron, refining of copper, refining of titanium and refining of tungsten are discussed due the their important applications of metals.

Extraction of Gold:

Gold is the precious metal and considered as rare metal. However, the gold availability in the ground is relatively larger than demand for gold. Gold is present as gold without combining any other elements and hence, gold is occurred as metallic state in the ore. Especially, gold is found with silver and hence, silver should be removed completely to have highest purity of gold.

Chemical methods are usually employed to extract gold. There are two chemical methods employed to extract gold and these are Gold amalgam method and Gold cyanide complex method.

Gold Amalgam Method:

Like other metals (silver and copper), gold forms compound with mercury when gold is made contact with mercury. This mercury compound of metal is in general called as amalgam compound. Hence, this method of extraction of gold using mercury carries amalgam name with it. When gold ore is treated with mercury, gold amalgam forms at the inner surface of drum that was used as a reaction container. Thus, gold is extracted as gold amalgam compound in the amalgam method.

Gold cyanide complex method:

This is other way of extraction of gold by reacting it with cyanide solution. Even though gold does not form compound with several known ligands, it readily forms compound with cyanide by formation of gold cyanide complex as shown below

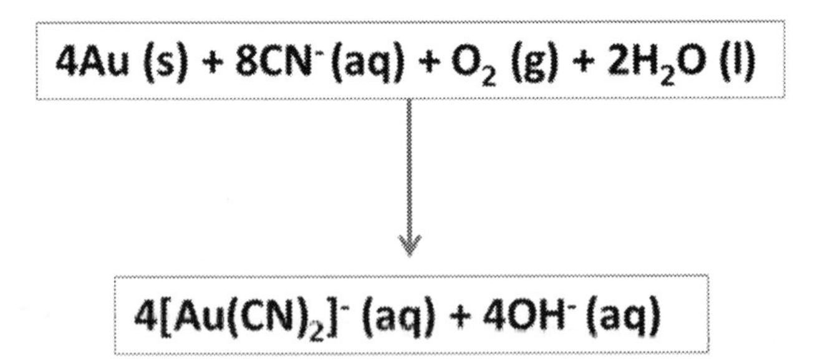

In this complex, gold is +1 oxidation state. Its tendency to form complex with cyanide ligand is due transition metal of gold. Oxide of gold such as Au_2O_2

is not stable but its +1 oxidation state with cyanide complex is the beauty to explore complex compound chemistry of gold to extract gold from its ore.

After having extracted gold as cyanide complex, it can be reduced back to gold by powdered zinc metal. Thus, zinc is oxidized into Zn^{2+} cation and it forms complex with cyanide while reducing Au^{+1} cyanide complex into gold metal. The chemical reaction involving redox reaction between gold (+1) cyanide complex and zinc metal is shown below in a chemical reaction.

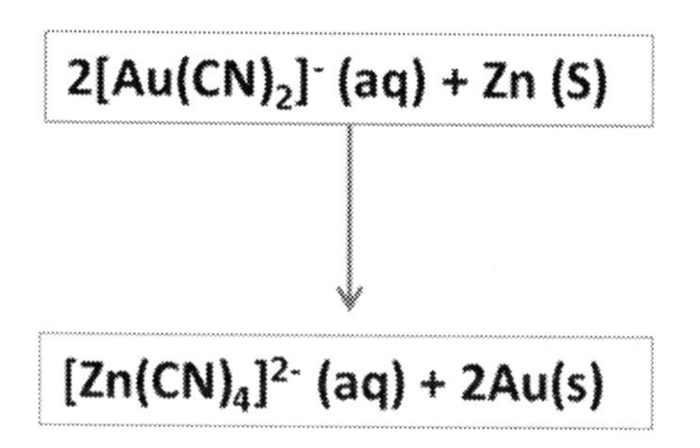

When silver and copper are present as impurities in gold ore, they can be easily separated before subjecting into cyanide complex method of extraction of gold process. Thus, chlorine gas is used as oxidant for silver and copper to remove from gold as AgCl and CuCl without affecting the gold medal. The formations of AgCl and CuCl are shown below in two separate chemical equations.

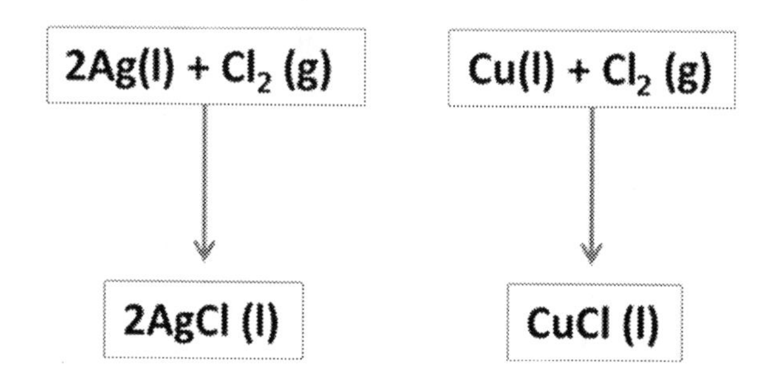

Refining of Iron:

Refining of iron refers to extraction and purification of iron from its ores. The principal ores of iron are

a. Hematite, Fe_2O_3, and hydrated Fe(III) oxides such as $2Fe_2O_3.3H_2O$
b. Magnentite, Fe_3O_4 with iron in its valences are +2 and +3.

c. Siderite, $FeCO_3$ (Iron (II) carbonate) and
d. Taconite (low grade iron ore due to presence of silica).

Majority of metal in iron ores is iron and they also consist of Silica, SiO_2 and other silicon compounds. Therefore, refining of iron from its ores is mostly involved reduction of iron cation into iron and purification of iron from SiO_2 and compounds of silicon. For this achievements, blast furnace is being used along with limestone, $CaCO_3$ and coke. Coke essentially produces thermal energy on combustion with oxygen as shown below at the the bottom of blast furnace.

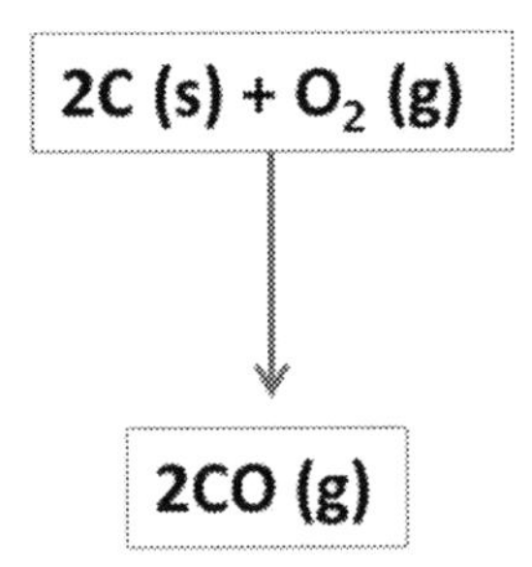

The carbon monoxide thus formed at the bottom of the blast furnace essentially used for the partial reduction of Iron (III) oxide into mixture of iron (II) oxide and iron (III) oxide, which is then followed by further reduction into iron metal through iron (II) oxide as shown sequence of several chemical redox reactions below.

1. Partial reduction of Fe^{3+} into Fe^{2+} so that the product consists of mixture of Fe^{3+} and Fe^{2+}

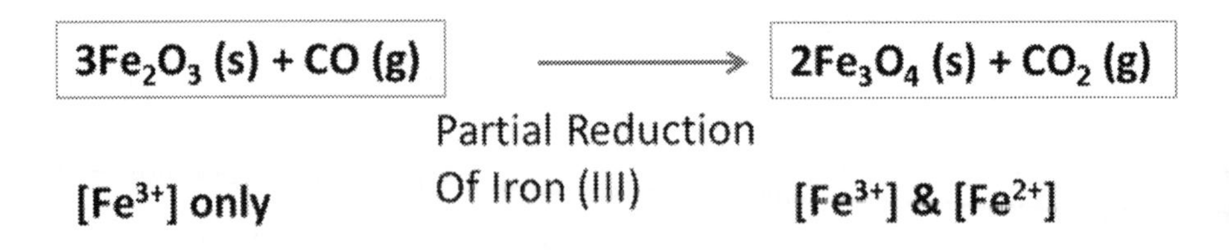

2. Further reduction of Fe^{3+} from the mixture of Fe^{3+} and Fe^{2+} into Fe^{2+} only.

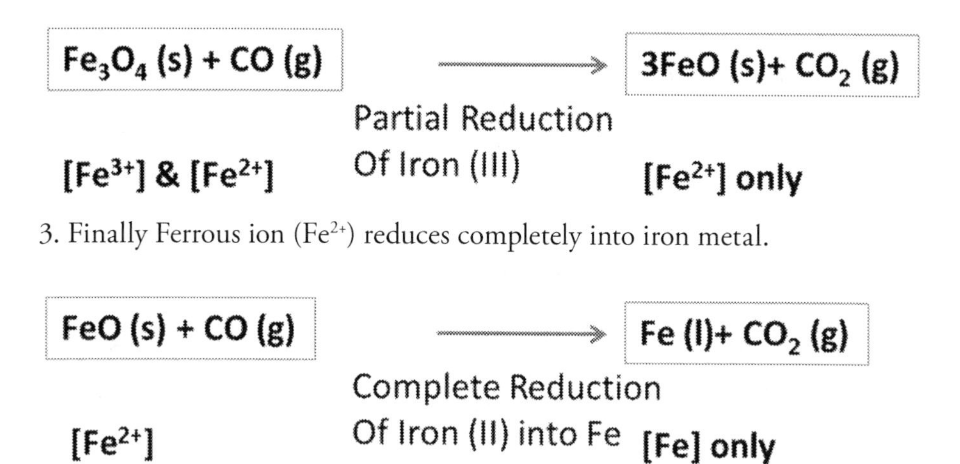

3. Finally Ferrous ion (Fe^{2+}) reduces completely into iron metal.

Interestingly, CO needed for the above stated reactions is usually regenerated by the reaction of CO_2 with coke as shown below.

The use of limestone, $CaCO_3$ in the refining of iron is separation of SiO_2 from iron ores by the following reaction between CaO (formed from $CaCO_3$ on decomposition) and SiO_2 as shown below.

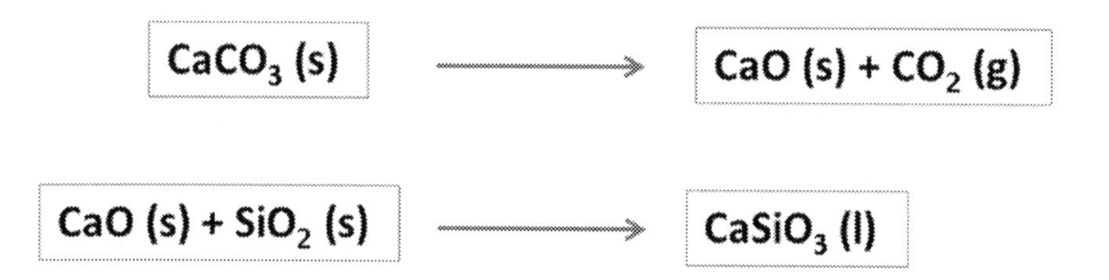

The calcium silicate as slag is less dense than the molten iron and hence, the slag floats on iron melt. Therefore, it can be drained off from time to time. Thus, iron metal produced from the blast furnace is so called cast iron or pig iron. The impurities present in the cast iron are

Carbon = 2 to 4.5%
Silicon = 0.7 to 3.0%

Sulfur = 0.1 to 0.3%
Phosphorous = 0 to 3.0 % and
Manganese = 0.2 to 1.0%

Therefore, the cast iron is relatively brittle in nature due to the presence of non-metallic impurities. The cast iron can be further purified by removing carbon and other non-metallic impurities, which ultimately lead to steel (alloy) with desired properties of flexibility, hardness, strength and malleability.

The cast iron thus further purified in an open breath furnace to burn off non-metallic impurities and this is usually achieved by adding calculated amount of iron (III) oxide. The reactions between iron (III) oxide and non-metallic impurities shown below are responsible for purification of cast iron into pure iron.

1. Removal of carbon as Carbon-di-oxide gas

$$3C\ (s) + 2Fe_2O_3\ (s) \longrightarrow 3CO_2\ (g) + 4Fe\ (l)$$

2. Removal of sulfur as Sulfur-di-oxide gas

$$3S\ (s) + 2Fe_2O_3\ (s) \longrightarrow 3SO_2\ (g) + 4Fe\ (l)$$

3. Removal of Phosphorous as P_4O_{10} gas

$$12P\ (s) + 10Fe_2O_3\ (s) \longrightarrow 3P_4O_{10}\ (g) + 2Fe\ (l)$$

4. Removal of Silicon as quartz liquid (SiO_2)

$$3Si\ (s) + 2Fe_2O_3\ (s) \longrightarrow 3SiO_2\ (l) + 4Fe\ (l)$$

CO_2 and SO_2 are gases and hence, they escape from the hearth open furnace. But, P_4O_{10} and SiO_2 should be removed from the atmosphere. Due to acidic nature, these non-metallic oxides, basic magnesium oxide or mixture of MgO – CaO is treated with acidic impurity oxides to remove them as salts. The reactions between acid and base are shown below.

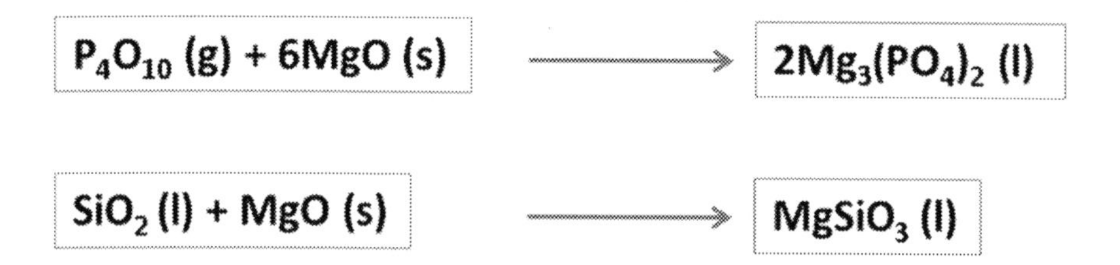

It it all sulfur reacts with iron to form FeS, which should also be removed due to its harmful impurity for steel making. The chemistry involved to remove FeS is by treating it with CaO (lime) and Coke as shown below.

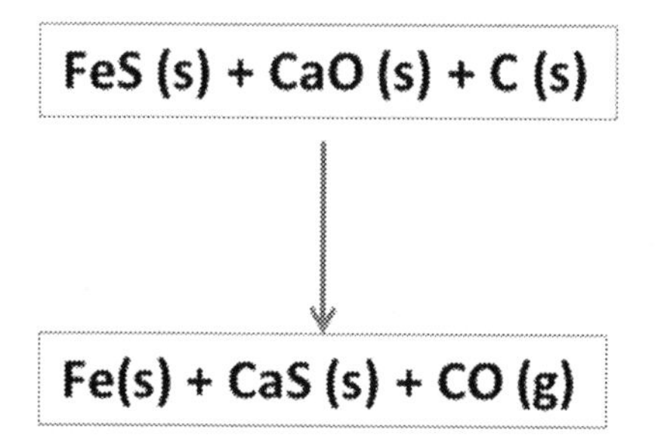

Refining of Copper:

Refining of copper refers to extraction and purification of copper from its ores. Copper occurs in the ground as two classes of ores and these are

a. Sulfide ores. They are $CuFeS_2$, Cu_3FeS_2 and Cu_2S
b. Oxidized ores. They are CuO, $Cu_2(OH)_2CO_3$ and $Cu_3(OH)_2(CO_3)_2$.

Refining of Copper from its sulfide ores:

It is important and essential to remove iron from sulfide ores of copper. The chemistry involved in the extraction (refining of copper) from sulfide ores are outlined now.

As usual, sulfide ores of copper are also subjected to purification, which then it is roasted in air at 1600°C. During the roasting of sulfide ores in air at such elevated temperature leads to useful chemical reactions and thus, arsenic and antimony are distilled off as their oxides while forming CuO, FeO and SO_2 gas. The chemical reaction is shown below.

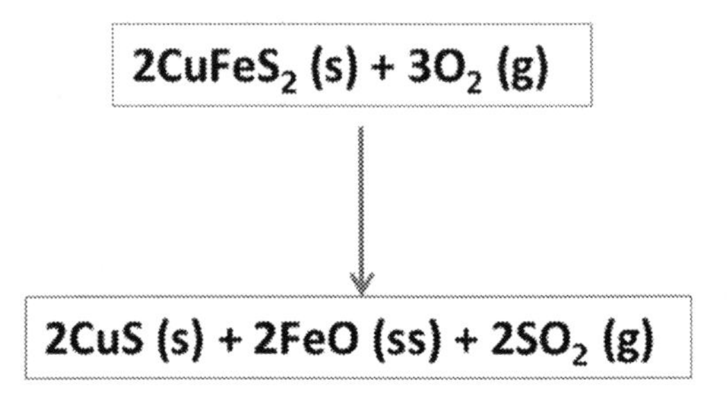

Now, the iron (II) oxide is removed using sand (SiO_2) and limestone ($CaCO_3$) in a so called reverberatory furnace. The chemical reaction involving FeO, sand and limestone is shown below.

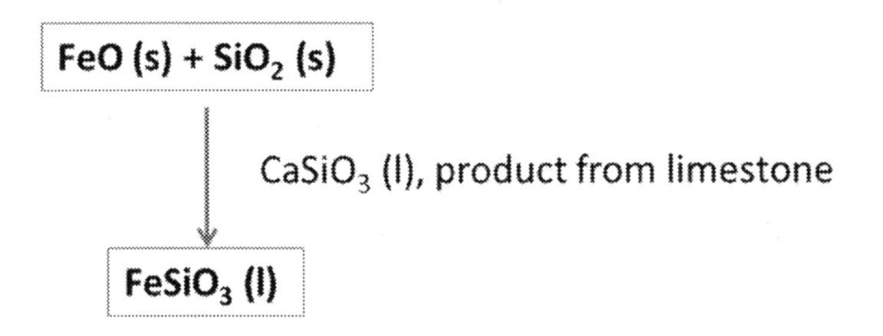

The iron silicate forms as slag and it is easy to separate.

CuS is converted into copper by the following chemical reaction.

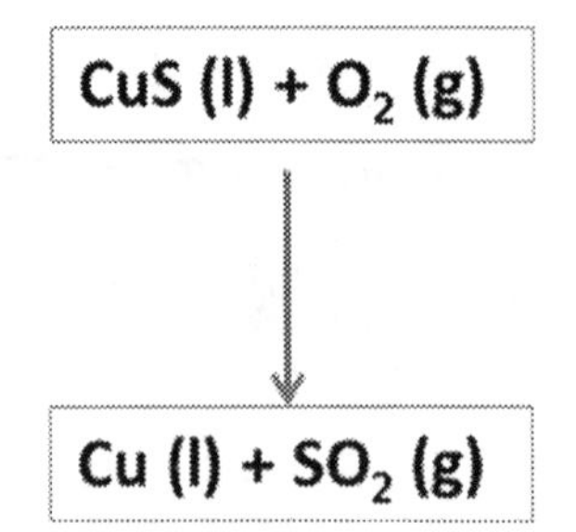

Or Cu_2S ore is directly converted into Cu by the similar way of converting CuS into Cu as shown below.

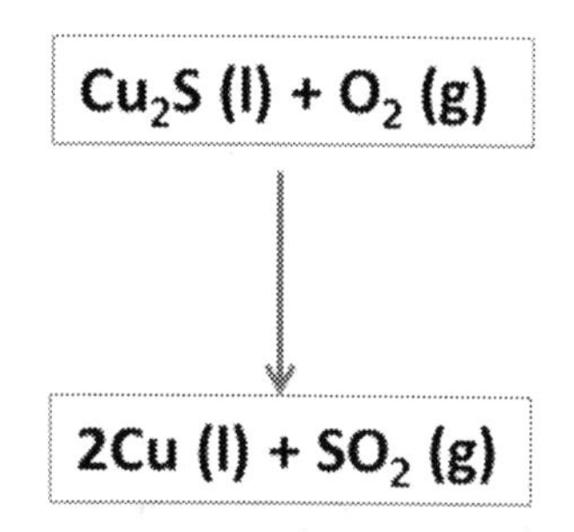

Copper finds a major application for electrical transmission and hence, it is understood importance of metallurgy of copper ores into copper metal.

Refining of Titanium:

Refining of titanium refers to extraction and recovery or purification of titanium from its ore. The chief ore of titanium is ilmenite, $FeTiO_3$ which is obtained along with silica. $FeTiO_3$ can be first separated from silica magnetically.

The chemistry involved in the recovery or extraction of titanium as $TiCl_4$ is the reaction among $FeTiO_3$, coke and Cl_2 gas as shown below.

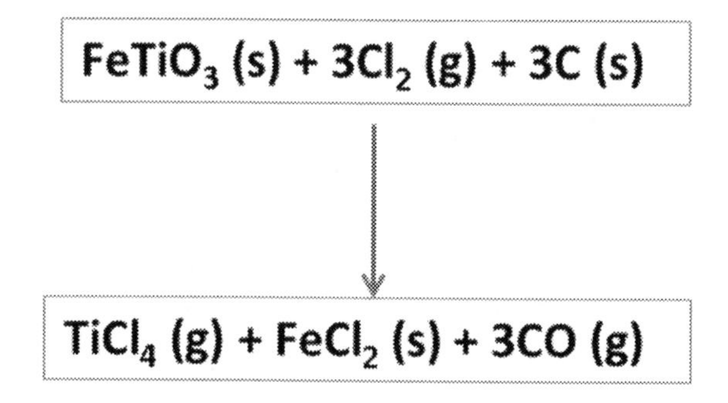

$TiCl_4$ is a volatile compound and hence, it can be purified by fractional distillation. Then, $TiCl_4$ is reduced into titanium by reactive metals such as Magnesium or Sodium as shown below.

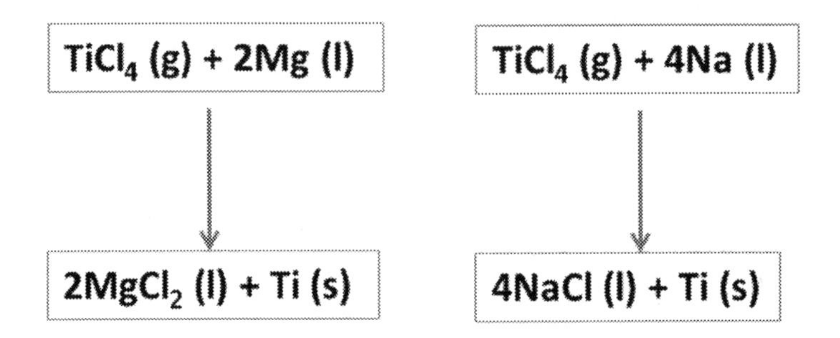

Water soluble side-products (NaCl and $MgCl_2$) in the both cases are easily washed out to recover titanium metal.

Properties and application of titanium:

Titanium metal has the following useful properties.

1. Tough metal
2. Low in density
3. Corrosion resistance and
4. Maintain high strength at high temperatures

Because of these unique properties of pure titanium, it has use in specialized aircraft engine and airframe application where cost is not an issue.

Refining of Tungsten:

Refining of tungsten refers to extraction or recovery and purification of tungsten. The main ores of tungsten are tungstates such as

1. $FeWO_4$
2. $MnWO_4$
3. $CaWO_4$ and
4. $PbWO_4$.

After initial purification into concentrated tungstates are treated with aqueous sodium hydroxide (NaOH) to produce tungstate as water soluble sodium tungstate, Na_2WO_4. The chemical reaction involving in the ionizable Na_2WO_4 is aqueous medium is shown below.

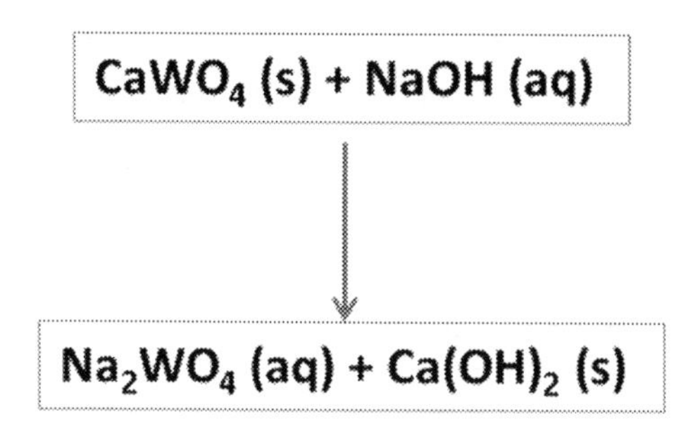

Then, Na_2WO_4 is treated with acid to recover tungsten as WO_3 as shown below in the chemical equation.

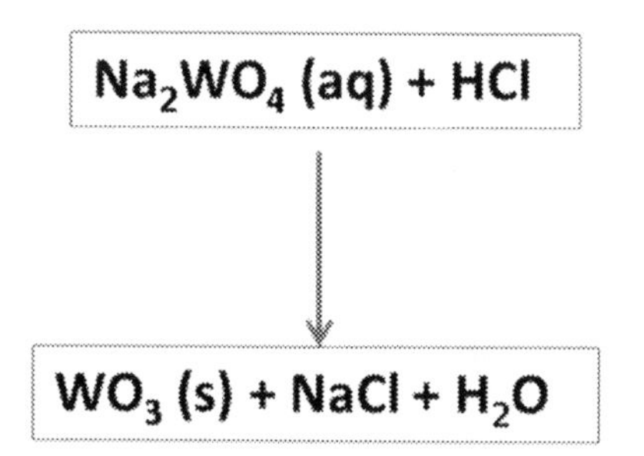

Finally, tungsten (VI) oxide, WO_3 is reduced into tungsten by treating it with hydrogen gas. The chemistry of reduction is shown below in a chemical equation.

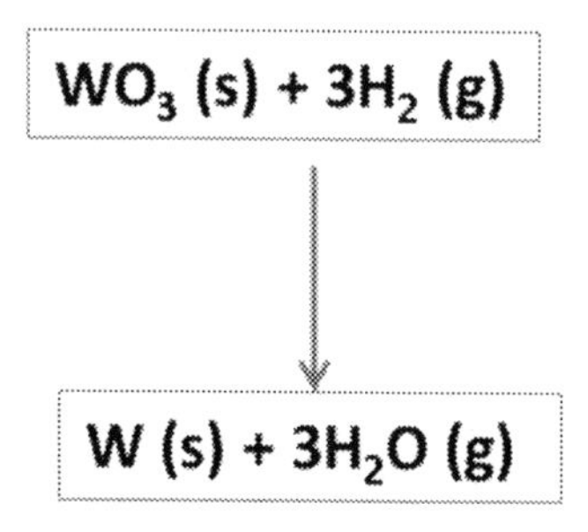

Properties and Applications of Tungsten:

Tungsten metal has the following useful properties.

1. Very dense metal (19.3 g/cm^3)
2. Extremely high melting point (3370°C)
3. High boiling point (5900°C) and
4. Very low vapor pressure even at high temperatures.

Tungsten metal finds useful in the following applications.

1. electric light filament
2. X-ray tube targets
3. Electrical contacts and arcing points
4. Furnaces
5. Making steel alloys, which help to have enormous increase in both hardness and strength.
6. Steels made from W and Cr find application in high speed cutting tools.

Powder Metallurgy:

Powder metallurgy consists of three major process sequence and these are

1. Blending and mixing of powders
2. Compaction and
3. Sintering.

Thus, powder metallurgy deals with powders of starting materials and it does not require wet or liquid chemistry. Flow chart 1 below outlines steps involved in the powder metallurgy process in general.

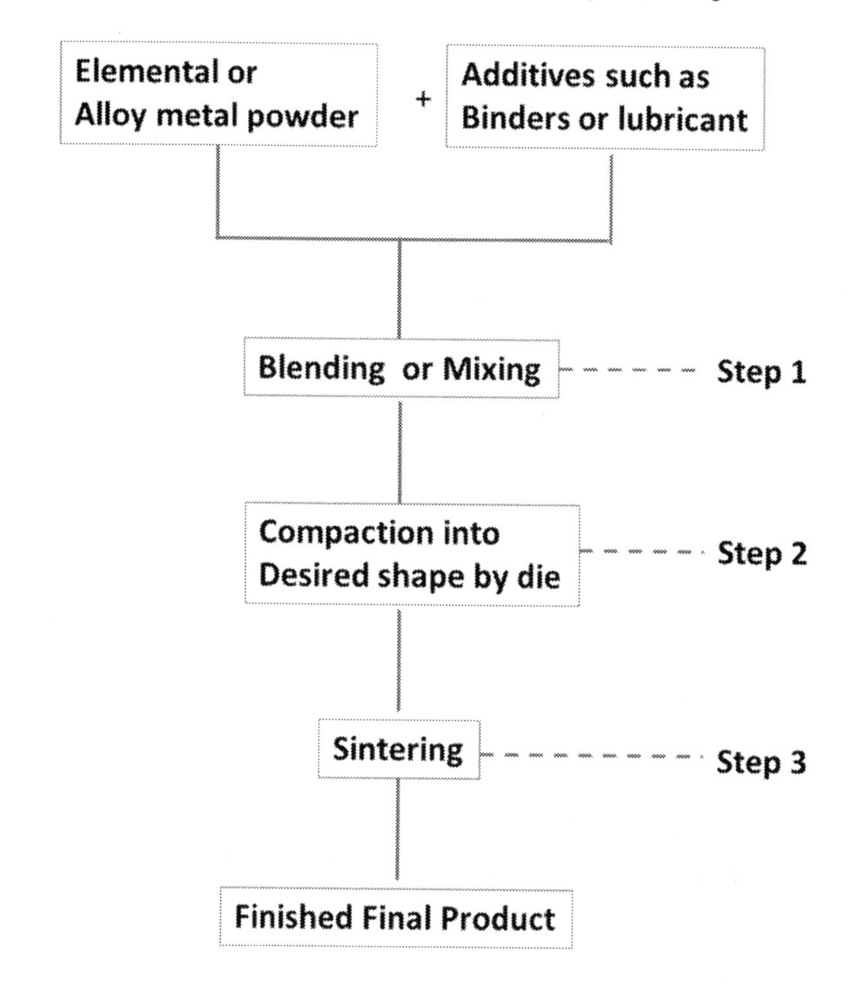

Flow chart 1: Outline of powder metallurgy

Step 1: Blending and Mixing:

Blending is the process by which mixing of powders with different sizes of the same chemical composition is achieved. Mixing refers to combining powders of having different characteristics.

Mostly in the blending and mixing process of powders lubricants, binders and deflocculates as ingredients are added. Each ingredient has its own function and role for better blending and mixing of powders. Thus, lubricants essentially reduce the friction among the particles. Binders are aimed to achieve enough strength to handle before sintering process. Deflocculates improve the flow characteristics during feeding.

Step 2: Compaction:

After having done with step 1 of blending and mixing of powders, the completely blended or mixed powders are pressed by high pressure using dies into the desired shapes. The desired shapes after compaction are called green compact or a simply green. The name given for the compact as green compact has meaning. According to my understanding it refers very similar to green vegetables (before cooking). Thus, before subjecting the compact to heating or annealing at elevated temperature is called green compact like green vegetables before cooking. Green compact can also be described as not fully processed compact or without sintering process in the step 3. Once the compact is sintered, it no longer carries the term green. Like vegetables after cooking it becomes curry. The green density of the compact has density that is usually greater than the density of blended powder. Density of green compact can also be improved either by application of double acting press instead of single punch or pressure is applied from all direction at room temperature (often called as cold isotactic pressing) or at higher than room temperature (so called hot isotactic pressing).

Step 3: Sintering:

Sintering refers to heating the green compact in a controlled atmosphere furnace to a temperature, just below the melting temperature of the powder but allow bonding of the particles. This step of sintering differentiates powder metallurgy from the regular metallurgy where roasting or melting is required. During sintering process, formation and growth of bonds between the particles occur.

Problems:

1. What is metallurgy?
2. What are the types of extractive metallurgy?
3. What are the steps involved in the pyrometallurgical process?
4. Describe the major steps of pyrometallurgical process
5. What is pyrometallurgical reduction process?
6. What is carbothermal reduction process?
7. What is metallothermal reduction process?
8. What is pyrometallurgical refining?

9. What is hydrometallurgical process?
10. What is the classification of hydrometallurgical process?
11. What is leaching?
12. Describe methods of purification metals
13. What is electrometallurgical process?
14. What is electro refining?
15. Describe extraction, properties and applications of Gold
16. Describe extraction, properties and applications of iron
17. Describe extraction, properties and applications of copper
18. Describe extraction, properties and applications of titanium
19. Describe extraction, properties and applications of Tungsten
20. Elaborate power metallurgy
21. What is sintering?

Topical Interest of Nano Materials

Objectives:

1. To outline two common methods of synthesis of nanomaterials
2. To explain co-precipitation method
3. To explain gas condensation method
4. To outline size dependent properties of nanomaterials
5. To outline size dependent physical properties
6. To outline size dependent structural properties
7. To outline size dependent electronic and phonon properties
8. To outline size dependent electrical and magnetic properties
9. To outline size dependent mechanical properties
10. To outline size dependent reactivity and catalysis
11. To outline size dependent optical properties
12. To explain quantum size effect
13. To study size effect on metal energy levels and semiconducting energy levels
14. To explain applications of quantum dots, graphene, inorganic phosphors, fuel cell materials (solid oxide and PEM fuel cells), thermoelectric materials, bioceramics and photo catalysts.

Nanomaterials are defined based upon their size that ranges from 0 to 100 nm in either 3-dimensional or 2-dimensional layered or 1-dimensional wire or zero dimensional dot. Thus, nanomaterials from a group of compounds which are neither atoms nor ions nor molecules not bulk systems. Therefore, these compounds have characteristic and distinct properties when they are compared with that of atoms or bulk compounds. This chapter covers general description of synthesis, size dependent properties, and exotic and emerging applications of nanomaterials.

Synthesis:

There are two common methods by which the synthesis of nanomaterials accomplished. One is called bottom-up method and another is called top-down method. The bottom-up method includes molecular or ionic precursor for the synthesis of nanomaterials. Whereas the top-down method is breakdown the bulk material into desired nanomaterial by mostly ball milling or physical gas phase evaporation or laser ablation method.
These methods are subject of interest with an example for each method of nanomaterials synthesis.

Bottom-up Method:

Co-precipitation Method:

Theoretical Principle:

This method involves simultaneously precipitating two or more metal salts to form intimate mixtures or solid solutions (precursor). Since water is used as solvent medium it is also called a wet-chemical method. The precursor is obtained either by adding a precipitating agent or by evaporating the solvent. The following equation represents the principle of co-precipitation as intimate mixture of two metal hydroxides.

$$MCl_2 + 2M'Cl_3 + H_2O \xrightarrow{\text{Addition of } OH^-} M(OH)_2 + 2M'(OH)_3 \text{ (precursor)} \quad (1)$$

$$M(OH)_2 + 2M'(OH)_3 \xrightarrow{\Delta} MM'_2O_4 \quad (2)$$

Since precursor is an intimate mixture of two metal ions, formation temperature of MM'_2O_4 is lowered when compare to that of solid state reaction between $M(OH)_2$ and $M'(OH)_3$.

Experimental Procedure:

Formation of $ZnFe_2O_4$ is explained in this section. Zn-oxalate and Fe-oxalate are dissolved in water. Then, water is evaporated to get precursor of oxalates of zinc and iron. Here, the precursor is a solid solution of zinc and iron oxalates. Then, this solid solution is heated to get powder of $ZnFe_2O_4$ compound. The following equation represents the formation of $ZnFe_2O_4$ powder.

$$Zn(C_2O_4) + Fe_2(C_2O_4)_3 + H_2O \xrightarrow{\text{Evaporation of water}} ZnFe_2(C_2O4)_4$$

$$\xrightarrow{\Delta} ZnFe_2O_4 + 4CO + 4CO_2. \quad (3)$$

The byproducts are gases, obtained solid product by this method is only $ZnFe_2O_4$. A disadvantage of this method is that precipitation of two or more cations, simultaneously, is critical and requires extreme care.

Top-down Method:

First and the foremost method is gas condensation method. It consists of vacuum chamber with pressure of $<10^{-5}$ Pa and it is equipped with a liquid nitrogen finger to collect metal nanoparticle and to scrapper assembly to scrap the metal nano particles from the wall of liquid nitrogen finger. Then, at the bottom of the chamber, nanoparticles can be collected. After having collected the nanoparticles by applying high pressure using piston at the bottom of the gas condensation apparatus nanoparticles can be compacted.

Size dependent Properties and their effect:

Nanomaterials have the following unique characteristics.

a. High Surface to Volume ratio
b. High Surface Area.

The size dependent properties and their effect are mainly due to the above mentioned two characteristics of nanomaterials. Thus, size dependency properties include

i. Thermodynamic Properties:

Surface Energy, Melting Point, Phase Transformations and Phase Equilibrium

ii. Physical Properties:

Phonon and electronic properties, electrical and thermal properties and magnetic properties.

iii. Mechanical Properties:

Superhardness, ductility and superplasticity.

iv. Chemical Properties:

Reactivity and Catalysis

v. Optical Properties:

Absorption and scattering of light.

The size dependent properties of nanomaterials are outlined one by one below.

Size dependent surface energy:

Concern over surface energy stems from the fact that estimation of characteristics of nanomaterials such as milling, dilution, wetting, nucleation, coagulation, recrystallization depends upon surface energy. For example, surface energy increases with decrease in particle size. This property, for example, affects the wetting property which is directly related to surface energy.

Size dependent melting point:

It is commonly observed that Nano size of material melts at lower temperature than that of of the bulk material. This is due to increase in atom thermal vibration amplitude at the surface layers. As a representative, melting point (Tm) of nano tin particles shows decrease in melting point with decrease in particle size (Depression in melting point)

Size Dependent Structure:

Particular structure formation is highly dependent upon size of the material. Thus, bulk ZrO_2 is monoclinic whereas nano ZrO_2 is tetragonal structure. Similarly, bulk $BaTiO_3$ has cubic structure whereas nano $BaTiO_3$ has tetragonal phase. Yet another example is that bulk TiO_2 is rutile structure but nano TiO_2 is anatase structure

Size dependent phonon and electronic Properties of Superconductivity:

As grain size decreases the superconductors transition temperature decreases. This is due to the decrease of the electronic state density at the Fermi level as the size decreases. Table 1 below summarizes grain size effect on superconductivity transition temperature, Tc of Nb film.

Table 1: Effect of Grain size on superconductivity temperature

Grain size, nm	Superconductivity transition temperature, Tc
60	9.4K
28	9.2K
17	7.2K
<8	<1.76K

Size dependent Electrical Properties:

It is clear and is known that the decrease in grain size is associated with the increase in conductivity and the decrease in the carrier activation energy. Table 2 below summarizes two examples to illustrate the size dependent electrical properties.

Table 2: Size Dependent electrical properties

Material	Grain size, nm	Temperature, K	Electrical Conductivity, ohm^{-1} cm^{-1}
Rutile TiO_2	50	713	$4x10^{-3}$
	260	713	$1.4x10^{-6}$
CeO_{2-x}	10	773	$1.6x10^{-5}$
	5000	773	$2.5x10^{-7}$

Size dependent Magnetic Properties:

Magnetic properties of materials are determined by strength of magnet that is usually measured and determined in terms of coercivity and magnetic saturation curve. The size dependent magnetic properties of coercivity and magnetic saturation indicates that coercivity decreases with decrease in the grain size whereas magnetic saturation increases with decrease in particle size for nano iron oxide particles. Table 3 below summarizes the size dependent magnetic properties of coercivity and magnetic saturation of magnetic particles of iron oxide.

Table 3: Size dependent magnetic properties

Average particle size, nm	Coercivity, Oe	Magnetic saturation at 11KOe, emu/g
60	107	62
51	106	77.6
43	88	79.5
35	82	77

Size Dependent mechanical properties:

The study of mechanical properties of metals, alloys, intermetallics, ceramics and polymers include strength, ductility, superplasticity, fracture toughness and thermo mechanical stability. Hardness of refractories increases by decreasing of grain size in the nanometer region of 0 to 100 nm. For example, the effects of grain size of a metal on its hardness and other properties are outlined in the Figure below. Thus, hardness of metal increases with decrease in grain size in the nanometer region and further decrease in grain size decreases the hardness. However, it is important to note that grain sliding phenomena takes place in the nanoscale grain size which is different from dislocation movement of grains at the micro/macro scale. Superplasticity of metals and alloys is usually defined as the ability of a material to exhibit a large degree of elongation prior to its failure, typically larger than 200%.

Size dependent reactivity and catalysis:

As particle size decreases the reactivity and catalysis increase. For example, the effect of Pt particle size on water gas shift reaction is outlined here. The water gas shift reaction is represented below in an equation.

$$H_2O + CO \rightarrow CO_2 + H_2$$

The water gas shift reaction is usually accelerated by Pt particles. Therefore, in order to understand its particle size effect on water gas shift reaction is an important example to consider here. The particle size of Pt nano particle decreases, the concentration of CO (one of the reactants) decreases. Thus, decreased size of catalyst Pt particles increases the water gas shift reaction.

Size dependent optical properties:

Optical properties of semiconducting metal oxides/sulfides (mostly) fall in the UV-Vis region and hence, it is interesting to consider the effect of particle size of semiconducting materials on absorption edge of them. In the case of semiconductors, nanoparticles with 1-20 nm have different optical/electrical properties from the bulk materials (>20 nm). An important note is that Nano ceramics has particle size of less than <100 nm whereas semiconductor nanoparticles have particle size in the order of 1-20 nm to exhibit different and interesting properties from corresponding bulk materials.

Electronic Energy levels of Semiconductor nanoparticles:

In the case of bulk semiconductors, band electronic structure exists whereas for molecular clusters with a small number of atoms exhibit discrete electronic structure. In the case of Nano semiconductors (number of atoms are between clusters and bulk) electronic band structure can neither have discrete cluster nor bulky band but in between them. The difference in electronic structure between bulk and nanocluster is shown in Fig. 1.

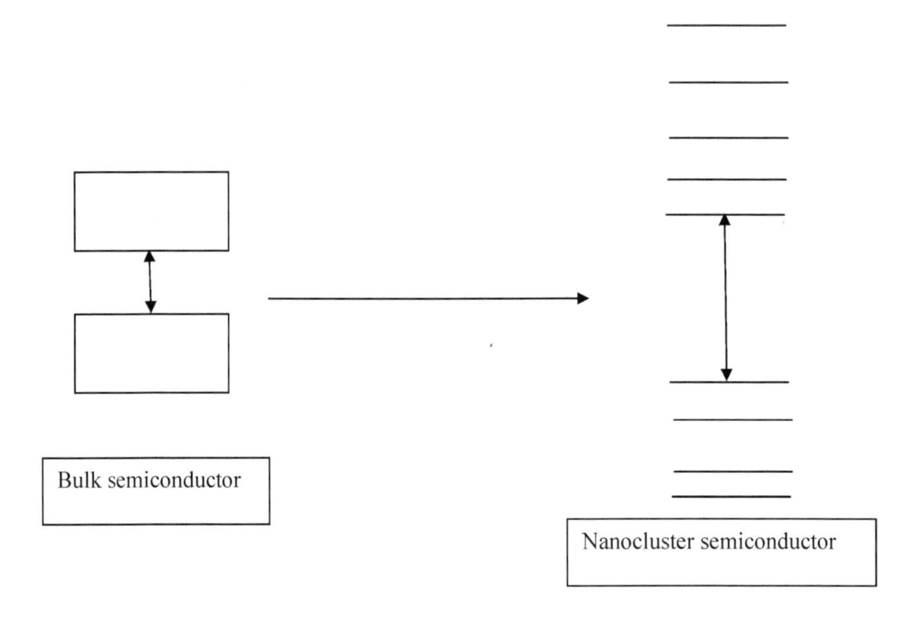

Fig.1: effect of decrease in particle size on energy levels

Nanocluster semiconductor has a larger band gap, as indicated by vertical two end arrows in addition to discrete electronic energy levels. Because of increase in band gap for nanocluster semiconductors, as shown above blue shift in the absorption edge is possible. In fact, experimental evidence reveals that in CdSe semiconductor as particle size decreases from 15 to 1.7nm absorption peak is blue shifted.

As particle size decreases the absorption peak of CdSe decreases to 400 nm and the body color of nanocluster CdSe with about 8 nm exhibits red color. Then, it turns to yellow to green to blue as size of particle decreases to about 2 nm.

Metal Nanoparticles:

Because of more surface to volume ratio, more atoms are present at the surface of nanoparticles. As surface atoms have unsaturated coordination environment due to their interaction with atoms inside the particles.

One of the important applications is the use of nano metal powder in electro catalytic PEM fuel cell. Because of requirement of Nano metal particles in this application, synthesis of stable Nano metal particle is a challenging. To get stable Nano metal particle, stable and high surface area inert support at which Nano metal powder is deposited to get, for examples, Pt or Pd at carbon electrode useful for fuel cell. The synthesis of Nano metal powder on the support is explained in the inorganic materials synthesis chapter as Electrostatic Adsorption Method. As stated in the method, Pt can be successfully deposited on carbon support with nano particle size. It is very clear that Pt nanoparticles (1 to 2 nm) obtained by Strong Electrostatic Adsorption method is smaller than that of Pt nanoparticles (8 to 10 nm) obtained by the conventional Dry Impregnation method.

Michael Faraday (1791-2867) discovered in the 19th century the unique optical properties of very small metal particles as various colors (ruby red to blue) of gold colloids with different sizes. The color of the small metal powder is due to colloid particle size, which is not observed for bulk Au metal. The absorption band observed for colloid Au metal is due to Plasmon resonance. Table 4 gives absorption maximum dependent on particle sizes of gold nanoparticles.

Table 4: Effect of Particle size on absorption maximum

Diameter, nm	Absorption Maximum, nm
8.9	517
14.8	520
21.7	521
48.3	533
99.3	575

The other important property of metals is electrical conductivity one. In the case of bulk metal electrical conductivity is observed and hence, bulk metal is metallic in nature. When size of metal is decreased, electrical conductivity is lost due to increase in gap between HOMO and LUMO and clusters in the range of a few nanometers show a transition from metallic to non-metallic behavior. This phenomenon is called size induced metal-insulator transition.

Selected Applications of Wonder Nanomaterials:

1. Quantum Dots:

Quantum dots are nanocrystals. They have practical-track applications depending upon their materials characteristics. Thus, quantum dots with electronic properties can find application as single electron transistors. Semiconductor quantum dots due to their atomic-like energy levels, particle size dependent fluorescence wavelength makes them in fabricating optical probes for biological and medical imaging. The other important applications are multifunctional creation from the assembly of receptor-ligand-mediated groupings and possible labeling of specific bioreceptors ligand mediated building blocks.

2. Graphene:

Graphene is from a layer of graphite. Below is the Fig.2 that compares between grapheme and graphite.

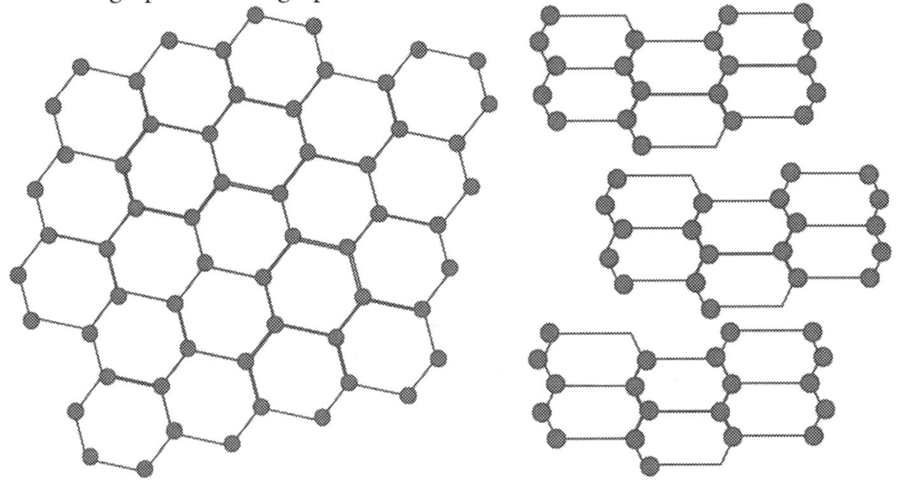

Fig.2 Structure of Graphene (left) and Graphite (right).

Due to one atom thickness of grapheme, unusual electronic spectrum is noticeably observed. Upsurge of research in graphene stems from the fact that its unique property in molecular electronics. It has extraordinary electric conductivity even at the limit of nominally zero carrier concentration and this observation is due to non-localized electrons in atom thickness of grapheme.

It indicates that graphene never stops conducting and in fact, conductivity in graphene is much more faster than electron in other semiconductors. Therefore, graphene finds immense applications in electronic industry. Other than electronic application of graphene, mechanical, thermal and optical property studies have been performed extensively. It has sensor application to detect dangerous molecules and also graphene sheet as membranes finds application in biomolecular sensors. Practical-track important application of grapheme is its composite materials with conducting polymer. It can also find application for thermal management in Nano electronics.

INORGANIC PHOSPHORS FOR SOLID STATE LIGHTING

Phosphors find various applications ranging from fluorescent lamp to immunoassay. Among them, fluorescent lamp is largely used and hence, lot of efforts is focused on it. Now, there is an immense research activity in academia and industries in inorganic phosphors after the discovery of blue or near UV LED. The combination of blue LED and yellow phosphor leads to replace the Hg discharge fluorescent lamp. Thus, toxic mercury could be successfully replaced now-a-days. Phosphors in the blue LED convert part of blue emission/ near UV of LED into green and red regions, and the combination of these colors results in white light. Nakamura patented yttrium aluminum garnet (YAG:Ce^{3+}) doped with Ce^{3+}, which is the best phosphor known today for obtaining white light from blue LED. Here, the part of blue light is transferred to yellow. Thus, the combination of blue and yellow leads to white light.

There are two known activators, Ce^{3+} and Eu^{2+} for the LED solid state lighting. These activators in a particular host exhibit required emissions. It is usually observed that Ce^{3+} doped phosphor shows emission about 100-150 nm lower than that of Eu^{2+} emission in the same host. Therefore, it is a thumb rule in the solid state lighting to predict the emission wavelengths for Ce^{3+} and Eu^{2+} activators.

Phosphors for Plasma Display Panels

When large screen with flat TV is attracted much interested today market in display companies, it is important to understand the function of it and the phosphors used in it.

There are four types of flat panel displays for a large screen display and these are display with cathode ray tubes (CRT), liquid crystal display (LCD), electroluminescence display (ELD) and plasma display panel (PDP).

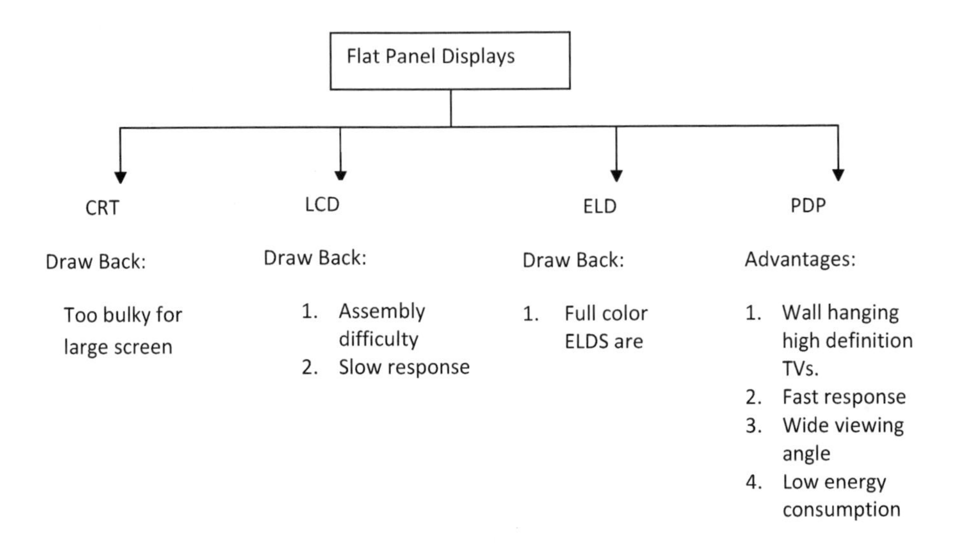

Scheme 2: Classifications of flat panel displays

From the Scheme 2 above it is very clear that PDP has several advantages over other displays for large screen display. Therefore, there is a large number of research activities on PDP phosphors to improve the existing phosphors or to discover potential phosphors.

Basic principle of plasma display panel:

This is made up of a glass substrate with two electrodes. Red, green and blue phosphors are coated inside of the glass substrate. Xe or Xe-He gas is filled with glass substrate. During discharge, the plasma produces shorter or higher energy wavelengths such as 130, 147 (the maximum intensity) and 172 nm. Phosphors coated inside the glass substrate convert high energy wavelengths into visible region.

PDP phosphors:

Currently used phosphors in PDP are Eu^{3+} activated Y_2O_3 and (Y,Gd) BO_3 red, Mn^{2+} activated Zn_2SiO_4 and $BaAl_{12}O_{19}$ green and Eu^{2+} activated

$BaMgAl_{14}O_{23}$ and $BaMgAl_{10}O_{17}$ blue phosphors. Thus, PDP phosphors resemble fluorescent lamp phosphors.

Characteristics of PDP phosphors:

1. Hosts and activators should be stable during fabrication and harsh operations of PDP.
2. Hosts and/or activators should absorb radiations between 130-180 nm and absorbed energy should be transferred into activators to show visible emissions.

Commonly used hosts for doping rare earth or transition metal ions are aluminates, borates, phosphates, silicates and some fluorides. Therefore, it is obvious that main group elements are only explored as hosts to get PDP phosphors. Even though, Mn^{2+} is well known green activator there is not much research on transition elements except Zn element. However, it may be possible to explore VO_4^{3-} group as host for PDP phosphors.

Rare earth ions such Eu^{3+}, Eu^{2+}, Tb^{3+} are explored as dopants in oxide hosts to get PDP phosphors. Table 5 summarizes decay times of commonly used PDP phosphors.

Table 5: Decay time of PDP phosphors

Phosphors	Decay Times, ms
Red: Y_2O_3:Eu	1.3
$(Y,Gd)BO_3$:Eu	4.3
Green: Zn_2SiO_4:Mn	11.9
$BaAl_{12}O_{19}$:Mn	7.1
Blue: $BaMgAl_{10}O_{17}$:Eu	<1
$BaMgAl_{14}O_{23}$:Eu	<1

It is very clear that when the decay times are almost same for blue and red phosphors it is very large for Mn^{2+} activated green phosphors.

Mechanism of PDP phosphors:

It is quite common that the higher energy of VUV radiation is absorbed by host lattice and it is transferred to activators to show visible emissions. In the borates, BO_3^{3-} group absorbs the 150 nm wavelength whereas 150-160 nm absorption in phosphates is due to PO_4^{3-} group. In the case of aluminates, broad excitation band near 175 nm is observed. Some fluorides such as $LiYF_4$, LaF_3 and YF_3 are having 120 nm excitation band.

Even though PDP phosphors resemble fluorescent phosphors, rare-earth doped $NaLaP_2O_7$ and $NaGdP_2O_7$ found useful in PDP cannot be used in fluorescent lamps because of Hg vapor reaction with Na ion present in the phosphors.

Fuel Cell Materials

Fuel cells operating at a range of temperatures will play an important role in the next generation of electricity production with zero pollution. Among the various fuel cells, solid oxide fuel cell (SOFC) and polymer electrolyte membrane (PEM) fuel cell are important to mention now. First, a fuel cell function is outlined and then, the main differences of the two fuel cells are highlighted.

Function of a fuel cell:

A fuel cell consists of anode, cathode and electrolyte. At the anode, fuel such as hydrogen gas is oxidized into proton and electron whereas at the cathode, oxygen is reduced into O^{2-}, which then combines with proton to yield water as gas. The electrons released at the anode passes through external wire to cathode and thus, electricity is produced. Function of a fuel cell is schematically described in Fig. 3

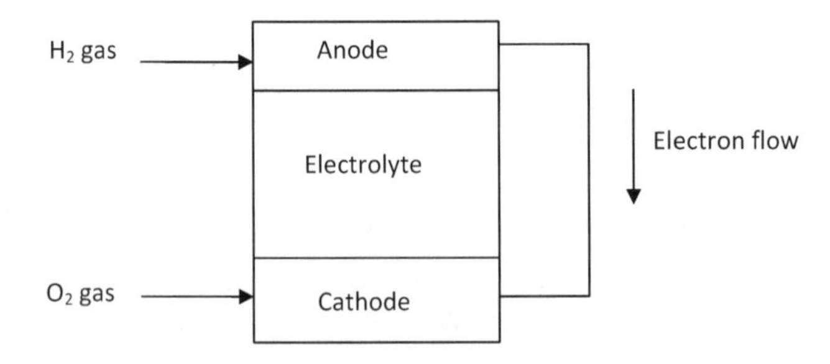

Fig. 3: Schematic diagram of a fuel cell

Brief description of PEM fuel cell:

Anode and cathode are platinum metals. Electrolyte is a polymer membrane. PEM fuel cell operates at 70°C and the main function of polymer electrolyte in the PEM fuel cell is to carry proton from anode to cathode, where in it combines with oxide ion to give water. Polymer membrane used currently is not stable beyond 100°C and hence, it is not possible to accelerate proton transportation from anode to cathode. Therefore, it is important to consider inorganic materials, according to me, containing proton as electrolyte. These inorganic materials might be stable enough to accelerate proton transportation from anode to cathode. Such examples are $H_4Nb_6O_{17}$, $H_2Ti_4O_9$, $HCa_2Nb_3O_{11}$ and $HNbWO_6$. Electrolyte should not allow the gases H_2 and O_2 to pass through it and there is another branch of materials science arises here to study the sintering property of protonated inorganic materials. Sintering is the process by which polycrystalline materials packed very closely such that polycrystalline materials achieve near 100% density of single crystal density. This is usually done below melting or decomposition point of the material. A schematic diagram showing function of PEM fuel cell is found in Fig. 4

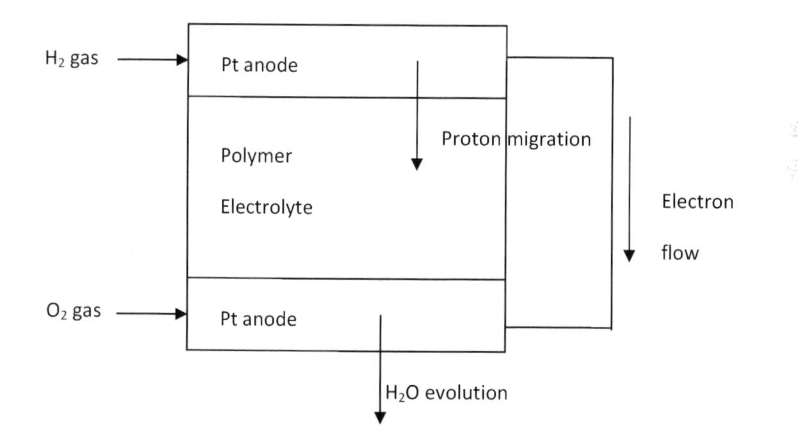

Fig. 4: Function of a PEM fuel cell

Brief description of SOFC:

In this fuel cell, anode is cermet and cathode is oxide. Electrolyte employed is oxides. The operating temperature is greater than 700°C. At this temperature oxide electrolyte transports oxide ion formed at cathode to anode where in oxide ion combines with proton to yield water. The required research in this topic is to reduce the operating temperature of SOFC. This is possible by investigating oxide ion transportation using various oxides. Fig. 5 shows a function of SOFC.

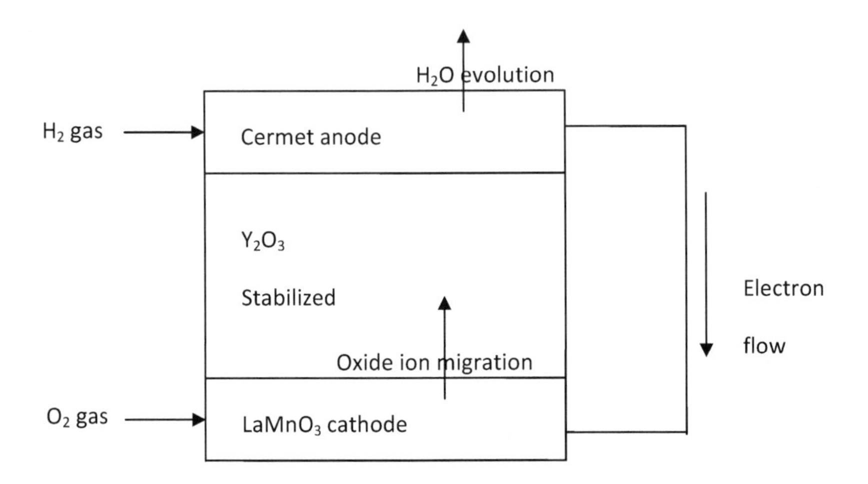

Fig. 5: Function of a SOFC

Differences between PEM fuel cell and SOFC:

Table 6 compares the main differences between PEM fuel cell and SOFC

Table 6: Comparison of PEM fuel cell and SOFC

PEM Fuel Cell	**SOFC**
Operating Temperature is ~80°C	>700°C
Polymer electrolyte	Oxide ion conductor as an electrolyte
Electrolyte carries proton from anode to cathode	Electrolyte carries oxide ion from cathode to anode
H_2O is formed at the cathode	H_2O and CO_2 are produced at anode
Electrodes are Pt	Anode is cermet and cathode is oxides

Thermoelectric Materials

The title itself implies that this material deals with thermal and electrical energies. Except material synthesis, all the properties are physics oriented. As a chemist I try to explain it very briefly in a simplified manner. Thermoelectric material is a semiconductor that finds applications in thermoelectric devices. These devices, making use of property of semiconductor, is simple, free from noise pollution and no toxic or no greenhouse gases evolve. Thermoelectric device converts thermal energy into electrical energy and as well as it cools one end when it is connected the other end to external input (Refrigeration).

Construction of Thermoelectric Device:

This device makes use of two different semiconductors, one is n-type and another is p-type semiconductors, connected in series. N-type semiconductor carries current by electron whereas p-type semiconductor carries current by hole. There are two applications found using these properties of semiconductors.

In a refrigeration, one end of two types of semiconductor is connected to external power. Charge carriers carry current as well as heat. Thus, external power attracts the charge carriers towards it. Because of movement of charge carriers towards external power, action of cooling takes place at the other end. This is due to carrying of heat by charge carriers as shown in the following figure (Fig.6).

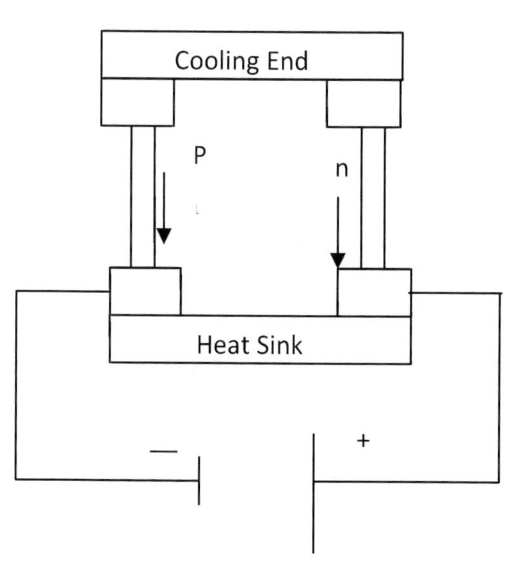

Fig. 6: Action of refrigeration using thermoelectric materials.

When one end of two semiconductors is heated voltage develops across the two semiconductors. Thus, heat is converted into electricity. When one end is heated, charge carriers move towards heating end. The following figure represents the function of power generation using thermoelectric materials (Fig. 7).

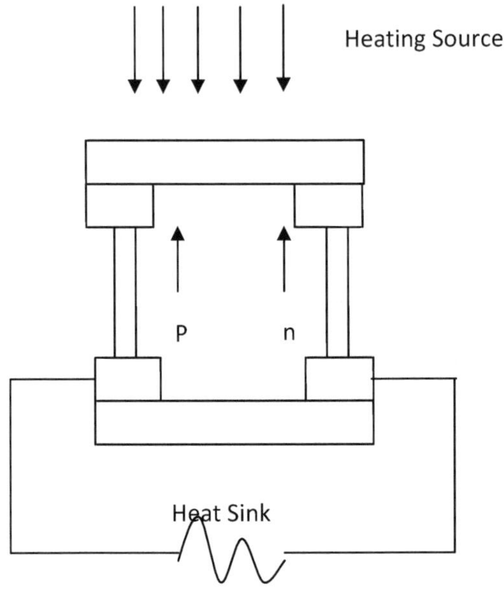

Fig. 7: Function of power generation using thermoelectric materials.

The efficiency of thermoelectric materials is determined by ZT where T= temperature and Z is a figure of merit. Z depends upon the electrical and thermal properties of semiconductors. To be a better semiconductor, they should show the lowest thermal conductivity and the highest electrical conductivity. Semiconductors having heavier elements are used in these applications since the thermal conductivity is lower for them. Recently, oxide materials are also found to show better property in these applications. A thermoelectric refrigerator requires ZT = 3 at room temperature.

Bioceramics

Bioceramics are mainly inorganic oxide materials and these find applications in replacing damaged and diseased part of the body. These are classified into two broad areas. One of them is bio inert materials that do not have any interactions with human body. Examples for this type of materials are α-alumina and zirconia. The another type of them is bioactive materials and they do have interactions with tissues in the body. Typical examples are calcium phosphate based oxides.

For the actual clinic applications, these bioceramics with good mechanical properties and chemical stability are required. For obtaining better mechanical property of these material, fine grain sintered body are prerequisite. Therefore, nanomaterials are required in this topic to get sinter-active materials.

It is also important to mention here that porous bioceramics are good for bone in-growth. However, macro porous materials as bioceramics are required but good mechanical property cannot be obtained with such a large porous. Therefore, in order to improve mechanical property, a secondary phase such as Al_2O_3, TiO_2, SiO_2, ZrO_2 is required.

Photo catalyst:

Solar-splitting of water using semiconductors plays a vital role in photo catalysis field. Using photon to split water into H_2 and O_2 is aimed at harness the solar energy because solar radiation reaching at the earth surface for 1 h is equivalent to fossil energy consumed by the world for 1 year. Therefore, research on photo catalytic breakdown of water is of worldwide interest. The reduction of water into H_2 and oxidation of water into O_2 using semiconductors are represented below.

$$2H_2O \xrightarrow[\text{Semiconductor}]{\text{Photon}} 2H_2 + O_2$$

The catalysts explored for photo splitting of water are called High Power Photo catalyst (HPP). However, catalysts employed to decompose or purify water or air are called Low Power Photo catalysts (LPP). These definitions of HPP and LPP are based upon my way of definitions and these would suite well in the field of Photo catalysts.

Principle of Photo Splitting of Water:

The principle of photo splitting of water into hydrogen and oxygen using semiconductors with appropriate band gap and positions of valence and conductance bands are important. The energy difference for the redox reaction of H_2O into H_2 and O_2 is 1.23 eV. Therefore, semiconductors with band gap greater than 1.23 eV are required for the solar-splitting of water into H_2 and O_2. Also, this photo evolution of gases from water depends upon the position of conduction and valence bands of semiconductors. Thus, the bottom level of the conduction band has to be more negative than the redox potential of H^+/H_2 (0 eV Vs NHE), while the top level of the valence band be more positive than the redox potential of O_2/H_2O (1.23 eV). It indicates that electron and hole are generated under photons with energy equal or greater than the band gap of semiconductor. Thus, electrons are promoted to the conductance band whereas holes remain in the valence band. These charge carriers are responsible for photo splitting of water. In a real situation, there are two main things occurring in a semiconductor particles

The desire thing is the charge separation of electron and hole and migration of them to the surface of the particle. In order to improve the catalytic activity, a co-catalyst such as Pt or NiO is coated at the surface of the particle. An unwanted process is recombination of charge carriers which lead to a lower yield for the reaction.

Among the semiconductors, TiO_2 is widely studied for photo splitting of water. TiO_2 semiconductor shows photo splitting of water into H_2 and O_2 under UV irradiation. Similarly, another system studied extensively is $K_4Nb_6O_{17}$. The success of $K_4Nb_6O_{17}$ semiconductor in the photo splitting of water is due the fact that hydrogen evolution and oxygen evolution take place in different interlayer space and hence, recombination of the gases is avoided. However, none of the layered compounds exhibited photo splitting of water into H_2 and O_2 under visible irradiation. On the other hand Pt/CdS is one of the photo catalysts that show evolution of H_2 gas using sacrificial reagent under visible irradiation, while WO_3 or $BiVO_4$ shows evolution of O_2 gas under the same conditions using different sacrificial reagent.

Questions:

1. What are the differences between bottom-up and top-down synthesis methods?
2. Explain size dependent properties with two examples.
3. State the size dependent catalysis reactions

4. What is quantum size effect and how does it affect the absorption edge of semiconductors?
5. What is graphene?
6. What are phosphors?
7. Compare SOFC and PEM fuel cell.
8. Explain action of refrigeration using thermoelectric materials
9. What are bioceramics?
10. Give two examples for photo catalysts.

Conventional and Non-conventional Energy Sources

Objectives:

1. To overview the energy sources and energy transformation needs.
2. To introduce conventional and non-conventional energy sources.
3. To outline the energy sources
4. To describe petroleum refining of thermal cracking process such as dewatering & desalting, atmospheric & vacuum distillation, thermal cracking, visbreaking & coking.
5. To show advantages of catalytic cracking over thermal cracking using fixed bed, moving bed and fluid bed catalytic crackings.
6. To outline additional petroleum refinery processes such as hydro treating, hydrocracking, thermal & catalytic reforming, isomerization, alkylation & polymerization processes, deasphating & dewaxing
7. To show end of conventional energy sources with natural gas in future
8. To describe alternate energy sources such as uses of Biomass, Bioethanol, Biodiesel for their needs and supplies.
9. To state the use of Hydrogen gas (H_2 gas) as an alternate energy source.
10. To define various fuel cells such as solid oxide fuel cell, polymer electrolyte membrane fuel cell, molten carbonate fuel cell and alkaline fuel cell.

Energy Sources and Energy Transformation:

There is no life without transformation of energy from one form to another form. However, conservation of energy states that energy neither created not destroyed. It can be transformed from one form to another form. Currently, more than 60% of energy supply is obtained from petroleum and natural gas on earth. In Asia, coal is a primary source of energy.

Classification of Energy sources as Conventional and Non-conventional:

Conventional energy sources are mainly from natural hydrocarbon products. They can be classified into Petroleum (fossil fuel) and Natural gas or gaseous fuels (fossil fuel). The conventional energy sources can't be replenished. Petroleum is essentially crude oil. Both petroleum and natural gas are, however, a mixture of hydrocarbons. At room temperature and atmospheric pressure, low molecular weight hydrocarbons such as methane, ethane, propane and butane are present as gases, while higher molecular weight hydrocarbons are in the form of liquids and/or solids. Both petroleum and natural gas are found in oil well and gas well respectively. Petroleum and natural gas from their well are not useful as such. Therefore, they are sent to refinery for purification and separation of different hydrocarbon molecules so that various compounds are of useful as fuels, lubricants, road asphalt and as feedstock for petrochemical processes to produce plastics, detergents, solvents, elastomers and fibers (nylon and polyesters). On refinery, crude oil yields three different groupings of products. The grouping is based upon their boiling point. Lower boiling point of hydrocarbons such as gas and gasoline form group 1 type. Group 2 type is middle boiling point of hydrocarbons such as kerosene, diesel fuel, fuel oil and light gas oil. Group 3 Type is remaining of crude oil with higher boiling lubricating oil, gas oil and residuum. Table 1 below summarizes different generic boiling fractions separated from a mixture of hydrocarbons that are present in crude petroleum oil.

Table 1: Fractions from crude petroleum

Group	Fraction	Boiling Range, °F
Low Boiling hydrocarbons	Light Naphtha	30 – 300
	Gasoline	30 – 355
Middle Boiling hydrocarbons	Heavy Naphtha	300 – 400
	Kerosene	400 – 500
	Light Gas Oil	400 – 600
	Heavy Gas Oil	600 – 800
High Boiling hydrocarbons	Lubricating Oil	>750
	Vacuum gas Oil	800 – 1100
	Residuum	>950

Non-conventional energy sources are also called as alternative fuels that are other than conventional energy sources. Examples include biodiesel, bio alcohol (methanol, ethanol and butanol), hydrogen and biomass. They can be solid, liquid and gaseous fuels. Biomass is of useful for direct heating or power, also known as biomass fuel. The alternate fuels can be replenished by planting.

Below is the outline 1 for the classifications of conventional and non-conventional energy sources.

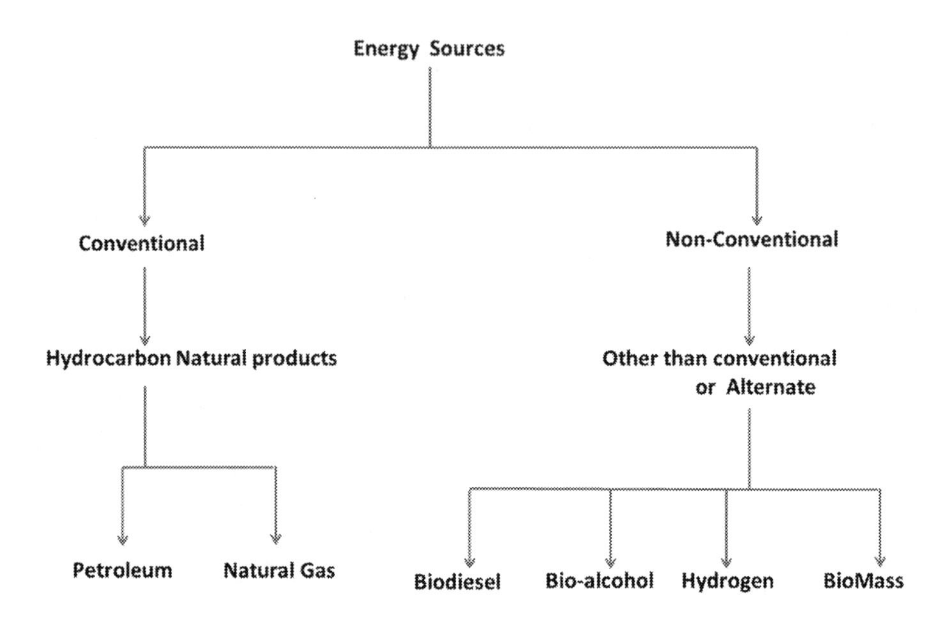

Outline 1: classification of Energy Sources

Petroleum Refining:

Dewatering and Desalting:

Petroleum from reservoir or well consists of gases, water and dirt (minerals). Therefore, the first step is to remove gases, water and minerals from crude oil.

The desalting refers to water washing operation for additional crude oil cleanup. The water washing step removes water soluble minerals that are present in crude oil. Electrostatic desalting is a common practice to remove water soluble minerals from crude oil. If the desalting step is not performed, then, the water soluble minerals can cause operating problem during refinery processing such as corrosion and catalyst deactivation.

Distillation:

There are two common distillation of refinery process. One is atmospheric distillation and another is vacuum distillation. Outline 2 below shows the two common distillation of refinery process.

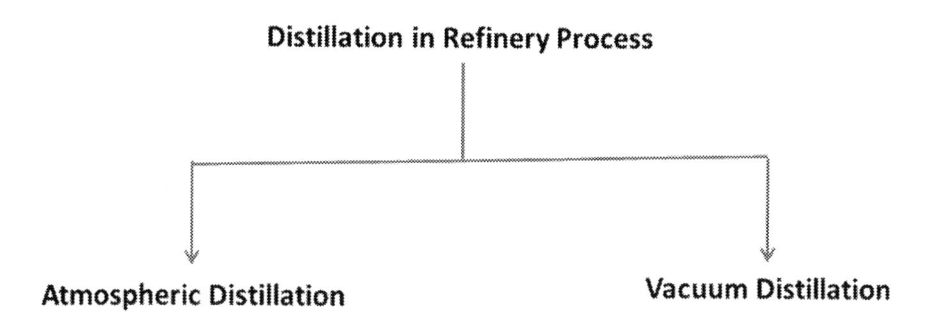

Outline 2: Two types of distillation in refinery process

Atmospheric distillation and vacuum distillation are briefly summarized one by one below.

Atmospheric distillation:

Atmospheric distillation explores tower unit for efficient degree of separation or fractionation. Feed is subjected to pipe still heater for pipe still furnace with predetermined temperature to heat the feed which is followed by entered into fractional distillation unit. At the furnace, the feed is vaporized and the vapor is held at the pipe under pressure until it is discharged into the fractional distillation tower as a foaming stream. The un-vaporized feed portion remains at the bottom of the tower. While the vapors pass through the tower to undergo fractional separation of gas oils, kerosene and naphtha. Thus, Pipe still furnace and fractional distillation units are the heart of the atmospheric distillation in the petroleum refinery. Fig.1 below illustrates the atmospheric distillation principles and primary separation of petroleum crude in the refinery process.

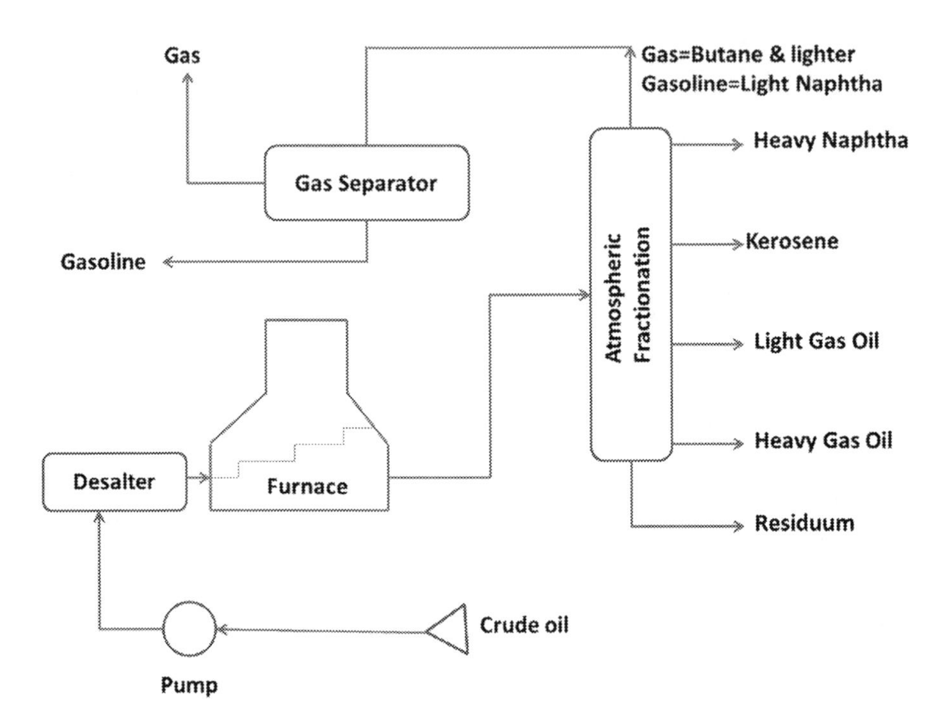

Fig.1 Outline of atmospheric distillation

In the atmospheric fractionation, from bottom to top heavy gas oil, light gas oil, kerosene and heavy naphtha are separated while residuum is recovered at the bottom. At the very top of atmospheric distillation Gasoline (light naphtha) and gases (butane and light) escape, which are subjected to gas separator to separate gasoline from other gases. Residuum from atmospheric distillation is a feed to vacuum distillation.

Vacuum Distillation:

To separate less volatile lubricating oils (products) vacuum distillation is explored in the petroleum industry refinery. The residuum from atmospheric distillation forms a feed to vacuum distillation. Vacuum distillation involves usually pressure of 50 to 100 mm Hg (atmospheric pressure = 760 mm Hg). Under this vacuum, heavy gas oil may be recovered at temperature 300°F and lubricating oil cuts may be obtained at 480 – 660°F. The vacuum distillation unit is outlined in the Fig.2 below.

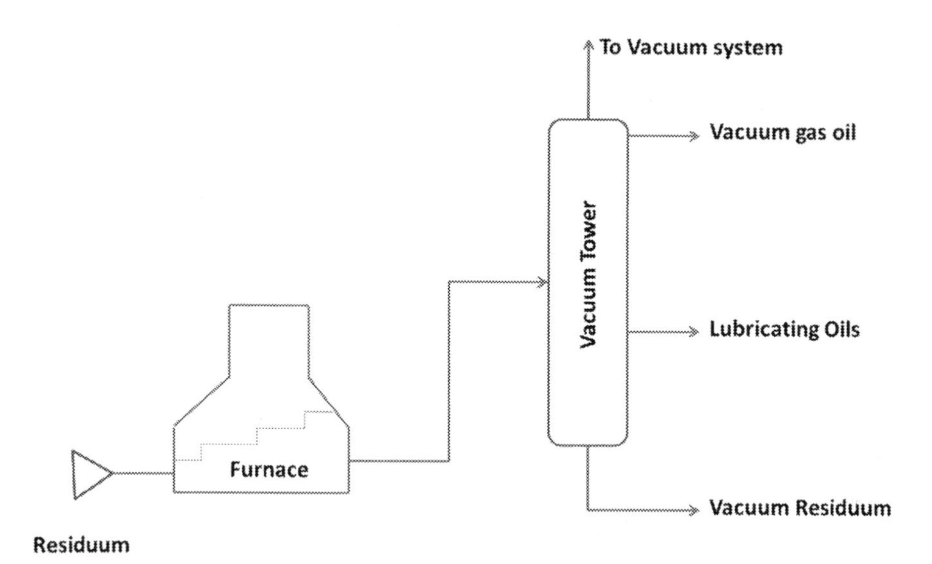

Fig. 2: Outline of Vacuum Distillation

It is noted that lubricating oils are separated from vacuum gas oil and however, vacuum distillation leaves vacuum residuum.

Thermal Processes:

There are three thermal processes carried out in the petroleum refinery industry and these are outlined in the outline 3 below.

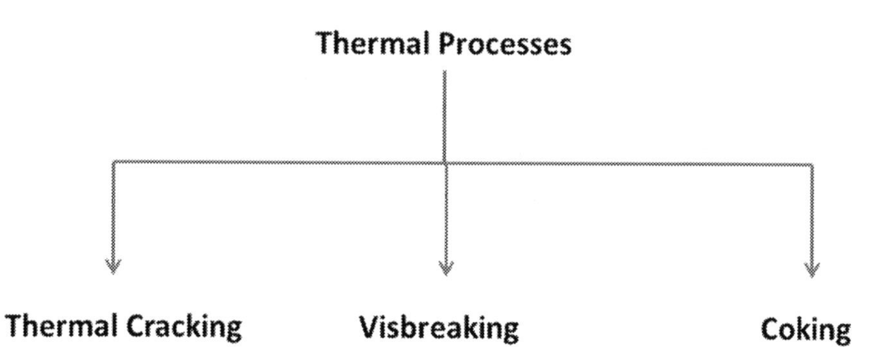

Outline 3: Outline of thermal processes in the Petroleum refinery industry.

These three thermal processes are explained one by one now.

Thermal Cracking:

Thermal cracking or cracking distillation usually refers to thermal decomposition of heavier nonvolatile materials into valuable lighter products or low boiling products. The valuable products are gas oils suitable for domestic and industrial fuel oils, and diesel. The gas oils are important source of gasoline. Thus, all the low boiling products obtained by thermal cracking could be used as such except that gas oils require further treatment to gasoline. Therefore, gas oils again fed along with fresh feed for going through cracking unit. Repeatedly performing the process yields more gasoline.

The temperature required for thermal cracking is in the range of 850 to 1005°F and pressures of 100 to 1000 psi. In the earlier days, the thermal cracking is typically called Dubbs process. In the Dubbs process, cracked gasoline and heating oil are recovered from the upper section of columns. Light and heavy distillate fractions are removed from the lower section.

Visbreaking:

Visbreaking refers to viscosity breaking by a mild thermal cracking operating. Thus, viscosity of residua to meet fuel oil specifications is achieved. Fig.3 below outlines the soaker visbreaker apparatus that is explored for viscosity breaking. Crude oil residuum as a feed is first heated to 895°F in a furnace. From the furnace outlet with its pressure of 100psi is then passed through soaking section. The charge remains in the soaking section until visbreaking reactions are completed. Then, mixture of cracked products is subjected to flash and fractional distillation chamber. Overhead product contains mainly low quality gasoline and light gas oil remains at the bottom. Liquid products from the flash and fractional distillation chamber are then sent to a vacuum fractionator to get heavy oil and residual tar of reduced viscosity.

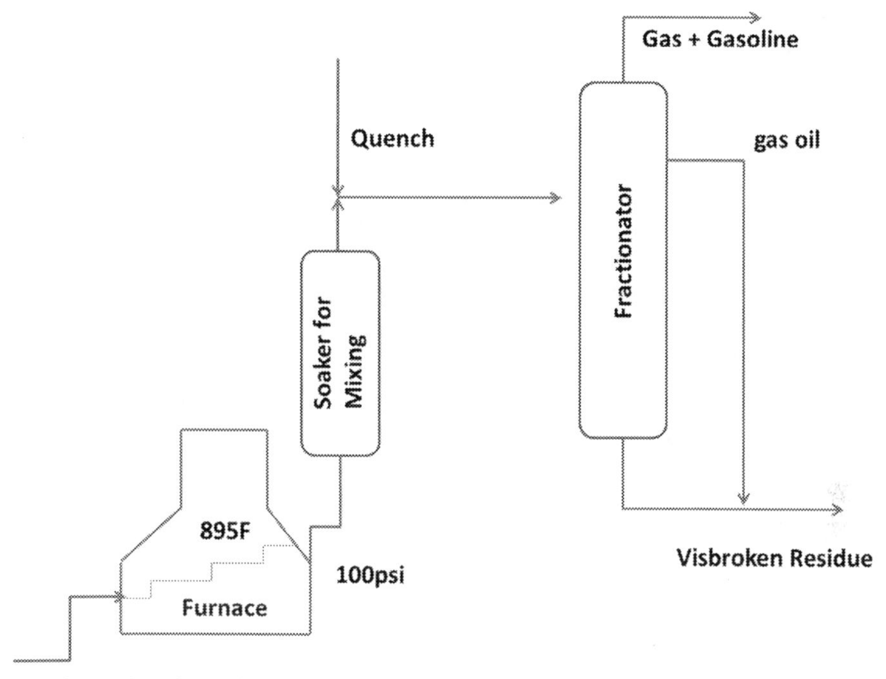

Fig. 3: Outline of Soaker Visbreaker Apparatus.

Coking:

Coking refers to conversion of heavy, low-grade oils (feed stock residuum) into light products and it involves complete thermal conversion into volatile products and coke. The volatile products include gases, naphtha, fuel oil and gas oil. With gas oil as feedstock, sulfur and metal impurities are indeed removed from coke. Coke finds useful as electrodes, production of chemicals and metallurgical coke. Typical examples for coking are delayed coking, fluid coking and flexicoking. Delayed coking is a semi continuous process, fluid coking and flexi coking are continuous processes. As a representative, flexi coking is outlined now.

Flexicoking:

Flexicoking was designed and made during the late of 1960s and the early 1970s. This method reduces excess coke and the heavier feedstocks in refinery operations. The heavier feedstocks are notorious for producing high yields of coke in thermal and catalytic operations.

After the coking process, feedstock of petroleum will be subjected catalytic cracking.

Catalytic Processes:

Advantages of catalytic cracking over thermal cracking process:

1. Gasoline with higher octane number is produced in the catalytic cracking.
2. Catalytic cracking yields higher octane number due to production of iso-paraffins and aromatics.
3. Catalytic cracking yields gasoline with low sulfur content.
4. Catalytic cracking yields more useful products than heavy residual or tar.

Over the years, methods developed for catalytic cracking are shown in the outline 4 below.

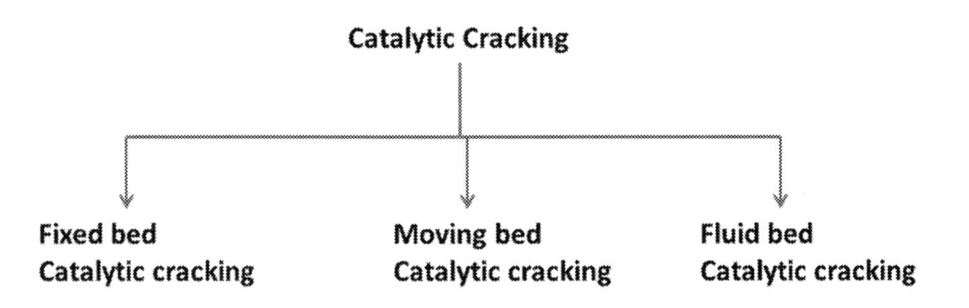

Outline 4: Three methods of catalytic cracking known so far.

Among these three methods, fluid bed catalytic cracking is very briefly outlined as a representative.

Fluid catalytic cracking (FCC):

Fluid catalytic cracking process involves catalyst powder that is circulated as a fluid with a feedstock. At the beginning of development of fluid catalytic cracking, natural clays have been explored as catalyst. Example for natural clay is Bentonite. Natural clays as catalyst existed until about 1936. Since then, silica-alumina has been explored as catalyst, which of course has better performance than natural clays. Activated catalysts also find useful in the fluid catalytic cracking. Activated catalysts are prepared by acid treatment. With the advent of microporous zeolites fluid catalytic cracking finds the microporous zeolites as proven and powerful catalyst. These zeolite catalysts are porous and hence, specific surface area is higher than the earlier catalysts. Therefore, highly adsorptive and their performance is significantly affected by the method of preparation. Thus, chemically identical catalysts, but different specific surface area due to different in pore sizes and shapes, they can have very different impact on fluid catalytic cracking such as activity, selectivity, reaction rates temperature and response to poisons.

Additional processes explored in the petroleum refinery industry are

1. Hydro processes
   - Hydro treating
   - Hydrocracking
2. Reforming Processes
   - Thermal reforming
   - Catalytic reforming
3. Isomerization processes
4. Alkylation processes
5. Polymerization processes
6. Deasphalting &
7. Dewaxing.

These seven processes are very briefly summarized below one by one.

1. Hydro processes:

Hydro processes involve thermal reaction of petroleum feedstock in hydrogen atmosphere so that coke formation is minimal and gasoline, kerosene and jet fuel are maxima. There are two main reactions that take place in hydro processes are Destructive hydrogenation and Non-destructive hydrogenation. Hydrogenolysis and hydrocracking are destructive hydrogenation. The hydrogenolysis and hydrocracking involve harsh reaction conditions. Simple hydrogenation is a non-destructive hydrogenation. This involves mild reaction conditions. The non-destructive hydrogenation process can remove nitrogen, sulfur and oxygen compounds into ammonia, hydrogen sulfide and water respectively so that pollution-free fuels are obtained. Table 2 below compares differences between hydro treating and hydrocracking processes.

Table 2: Comparison between hydro treating and hydrocracking processes

Hydro treating	**Hydrocracking**
Mild conditions with H_2 atmosphere are used	Harsh conditions with H_2 atmosphere are required
Non-destructive process with removal of sulfur, nitrogen and oxygen compounds that are present in the feedstock	Destructive process with hydrogenolysis and hydrocracking
Examples for hydro treating catalysts include Co-Mo or Ni-Mo loaded alumina	
Hydrogenation of aromatic does not take place.	Here, coke formation and polymerization are minimal
Temperature ~700°F; H_2 pressure ~500-1000 psi	Temperature ~800°F, H_2 pressure=1000 psi
Desulfurization is achieved with small amounts of hydrogenation and hydrocracking	-

Hydro treating:

Hydro treating refers to reactions of feedstock with hydrogen in presence of catalyst. The catalyst should be stable enough to sustain in the hydrogen atmosphere and should accelerate the required reactions. Examples for hydro treating catalyst are tungsten-nickel sulfide, cobalt-molybdenum-alumina, Nickel oxide-silica-alumina, alumina supported Pt and Molybdenum-cobalt catalyst. The temperature of the reactions ranges from 500 to 655°F and hydrogen pressure ranges from 500 to 1000 psi.

Hydrocracking:

Hydrocracking is very similar to thermal cracking except that hydrogen atmosphere is introduced in the hydrocracking. The feedstock that is difficult to process either by catalytic cracking or by reforming to upgrade low-value distillate such as cycle oils, coker gas oils and heavy cracked and straight non-naphtha is accomplished in the hydrocracking process, as added advantage of destructive hydrocracking process.

Thus, feedstock is heated and passed along with Hydrogen gas into contact with catalyst pellets. The temperature of reaction is between 500 and 800°F at hydrogen pressure that ranges from 100 to 1000 psi. The actual temperature range and pressure range are determined by particular process that is required for the nature of feedstock and degree of hydrogenation. After completion of expected reactions in the hydrocracking reactor, hydrogen is separated and recycled through the reactor by removal of H_2S. Then, H_2S removal is required at the stripping tower. Finally, the product is cooled down and stored or sent for the next treatment steps.

Reforming Processes:

As the demand for higher octane gasoline has been asked during the early 1930s attempt was was developed to improve octane number of fractions within boiling point of gasoline by a process that is so called reforming process. The reforming process is widely used but lesser extent than thermal cracking. Thus, reforming process converts (reforms) gasoline into higher octane gasoline. There are two reforming processes known. One is thermal reforming and another is catalytic reforming. Table 3 below summarizes comparison between thermal and catalytic reforming.

Table: 3 Comparisons between Thermal and Catalytic Reforming

Thermal Reforming	**Catalytic Reforming**
Products are gaes, gasoline and residual oil or tar	Dehydrogenation is a main chemical reaction in catalytic reforming and hence, more H2 gas is formed.
At higher temperature, gasoline with higher octane number is obtained.	Composition of reforming catalyst is dictated by the composition of the feedstock and desired reformate

Reformate with octane numbers of 65 to 80 is obtained	Reformate with octane numbers of 90 to 95 is achieved.
At higher temperature, yield of gasoline with higher octane number is low	Sulfur can poison the catalyst
Less effective and less economical	Highly effective and economical

Thermal Reforming:

Feed for thermal reforming process is naphtha or a straight run gasoline. If naphtha is a feed, the temperature of furnace is 400°F and if a straight run gasoline is a feed, the temperature of furnace is 950 – 1100°F. Pressure of the furnace ranges from 27 to 68 atmosphere. During the reforming reactions in the furnace at high temperature and high pressure, naphtha is passed into fractional tower to separate heavy products. After the reforming reactions in the furnace and separation of heavy products in fractional tower, higher octane olefins are obtained as a primary sources of reformate. Thus, products in the thermal reforming process include gases, gasoline and residual oil or tar with minimum quantity of residual oil or tar.

The amount and quality of gasoline as reformate significantly depends upon the temperature of furnace. The thumb rule is that with increase in temperature of furnace, octane number of gasoline increases, but at the expense of gasoline (reformate) yield. The octane number of gasoline achieved in the thermal reforming process is only 65 to 80.

By and large, thermal reforming process to get higher octane number gasoline is less effective and less economical when it is compared with that of catalytic reforming process.

Catalytic Reforming:

Catalytic reforming refers to conversion of low octane gasoline into higher octane gasoline as reformate. However, catalytic reforming achieves octane number of gasoline as 90 to 95. The role of catalyst in catalytic reforming is hydrogenation – dehydrogenation catalyst and it is usually supported on alumina or Silica-alumina.

There are three main types of commercial catalytic reforming processes known and these are moving-bed, fluid-bed and fixed-bed types. Catalyst employed in the catalytic reforming process depends upon the type of process.

Thus, moving-bed and fluid-bed processes explore mixed metal oxide catalysts whereas fixed bed type catalytic reforming process demands precious metal like Platinum as catalyst.

Feed for catalytic reforming process is straight run naphtha and hydrocracker naphtha. In the catalytic reforming process, a prime reaction of dehydrogenation takes place and hence, lot of hydrogen gas is produced. The hydrogen gas in fact is recycled for hydrogen atmosphere in the furnace which is of course consumed during catalytic reforming process.

Composition of feedstock and desire of reformate determines the composition of reforming catalyst. Thus, catalysts employed in the reforming process are MoO_2-Al_2O_3, Cr_2O_3-Al_2O_3 (mixed metal oxides) or Pt loaded at the surface of SiO_2-Al_2O_3 or Al_2O_3 support. In order to eliminate the problem of poisoning of Pt by sulfur gas, mixed metal oxide catalysts are proven to be beneficial in the catalytic reforming process.

Isomerization Processes:

High octane gasoline from heavier gasoline fraction is being achieved by catalytic reforming. But, isomerization processes refer to conversion of lighter gasoline fraction especially butanes, pentanes and hexanes into their isomers that have high octane rating in their lower boiling range. Thus, the straight-chain paraffins (such as n-butane, n-pentane and n-hexane) are converted into iso-compounds respectively. The isomerization processes are achieved by activated aluminum chloride catalyst with as low as temperature possible. The other function of the activated aluminum chloride catalyst (first ever reported catalyst) is to prevent side reactions of cracking and olefin formation. However, Platinium supported catalysts have also found for use in high temperature of operations in the range 700 to 900°F with 20 to 51 atm pressure.

Alkylation Processes:

Alkylation processes refer to formation of higher iso-paraffins by alkylation of olefins with paraffins. If olefins are not alkylated by this process, their presence in the gasoline might yield pollutants in the automobile exhaust. Therefore, iso-paraffins with high octane gasoline is highly preferred, which is of course being accomplished by alkylation processes. For alkylation processes, acid catalysts such as aluminum chloride, sulfuric acid or hydrogen fluoride are explored in the commercial alkylation processes.

Polymerization Processes:

Polymerization processes refer to conversion of olefin gases into liquid products of gasoline (also known as polymer gasoline). The polymerization processes are achieved either thermally or in the presence of a catalyst at lower temperature. As usual, thermal process is not effective and less economical due to its high temperature operation and poor yield. Acid catalysts are known for effective polymerization of olefins into polymer gasoline. Examples for acid catalyst are sulfuric acid, copper pyrophosphate, phosphoric acid which works even at 300 – 425°F and 10 to 81 atm pressure depending upon the nature of feedstocks and desired products. Phosphates are known catalysts for polymerization process.

Deasphalting:

Major part of refinery operations is solvent deasphalting processes. In the deasphalting processes, precipitation of polar constituents takes place. When propane is used for the deasphalting processes, precipitation as deasphalted oil (DAO) and propane deasphalter asphalt (PD tar) takes place. It is known that paraffins are very soluble in propane at normal temperature but at higher temperature (200°F) all hydrocarbons are almost insoluble in propane.

Dewaxing:

Dewaxing refers to removal of paraffin waxes from the crude oil. To achieve this, methyl-ethyl-ketone (MEK) is used as solvent to remove this was before it is processes. Thus, dewaxing involves solvent and however, it is not a quite common practice.

Natural Gas:

It is a fossil resource. It can be used as an energy source in power plants or as a heating source. It also forms a major basis for its transformation into many basic materials for chemicals. In the industry, its conversion to syngas via steam reforming, partial oxidation or autothermal oxidation is practical. Conversion of natural gas with possible reactions is summarized below.

a. Steam Reforming:

1. $CH_4 + H_2O \rightarrow CO + H_2$
2. $CO + H_2O \rightarrow CO_2 + H_2$ (water gas shift reaction)

b. Dry (CO_2) Reforming:

3. $CH_4 + CO_2 \rightarrow 2CO + 2H_2$

c. Autothermal Reforming (ATR):

4. $CH_4 + 0.5O_2 \rightarrow CO + 2H_2$
5. $CH_4 + H_2O \rightarrow CO + 3H_2$
6. $CO + H_2O \rightarrow CO_2 + H_2$

d. Catalytic partial Oxidation:

7. $CH_4 + 0.5O_2 \rightarrow CO + 2H_2$

e. Total Oxidation

8. $CH_4 + 2O_2 \rightarrow CO_2 + 2H_2O$

The chemical products that are reasonably obtained from natural gas are summarized below in Table 4.

Table 4: Chemicals Products from natural gas

Products	**Main Technology**
Ammonia	Syngas/synthesis
Ethylene	Steam cracking of C_2H_6
Propylene	Steam cracking
Methanol	Syngas/synthesis
Hydrogen	Steam Reforming
Synfuel	Syngas/synthesis

Calorific Value of Fuel:

Calorific value of fuel is defined as the total quantity of heat generated on burning a unit mass of fuel completely. Its unit is cal/g or kcal/g or Btu/lb. The calorific value of fuel is estimated in a bomb calorimeter under constant volume conditions. Thus, a known amount of fuel is burnt in bomb calorimeter to know its effect on rise in temperature of water surrounding the calorimeter indicates amount of heat released by the fuel under constant volume.

Calorific Intensity:

When a fuel is subjected to complete combustion the maximum temperature obtained is called calorific intensity.

Non-conventional Energy Sources or Alternate Energy sources:

In this section, main focus will be biomass and hydrogen fuel cells (as representative) as alternate energy sources, which are of course renewable.

Biomass:

Biomass refers to non-animal renewable resources such as trees and crops that can be converted to harvest for energy or as chemical feedstock. Biomass is a small amount when it is compared with that of fossil resources, but its renewable nature makes it attractive. It is important to understand energy from biomass by various methods that are known until now. Like coal, it can be burnt out to get heat energy or electricity in the power plant. Since it is a complex mixture of starch, cellulose etc., its burning process is as shown below.

$$[CH_2O] + O_2 \rightarrow CO_2 + H_2O$$

During this burning process, lot of energy is released. Also, it is quite important to note that CO_2 release is also more. However, CO_2 is consumed in growing the biomass. However, biomass needs to be compete with fossil fuels and therefore, primary conversion processes developed so far are summarized in the Fig.4 below.

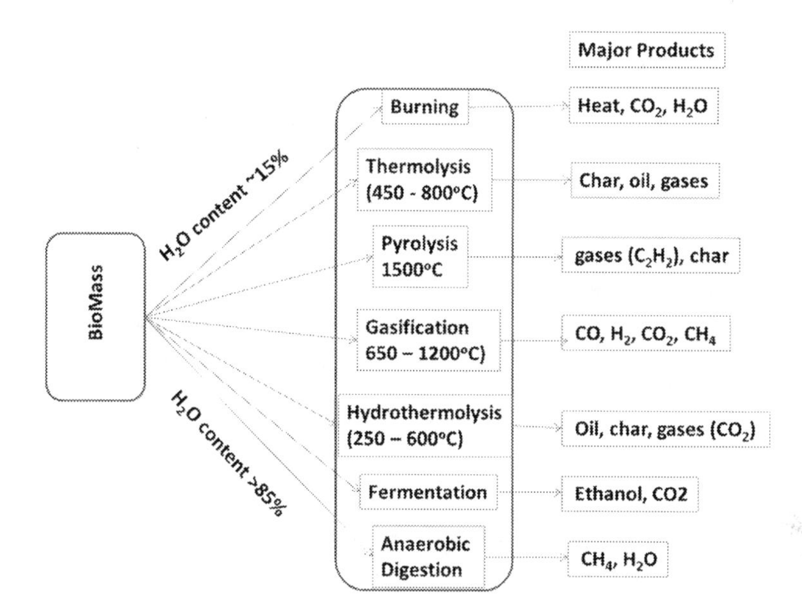

Fig.4: Possible products by different routes from BioMass.

Biomass conversion processes are outlined very briefly one by one.

Thermolysis and Pyrolysis of Biomass:

These two processes involve breaking the long chain gigantic molecules mostly in the absence of oxygen gas at temperature 450°C< >800°C for thermolysis and about 1500°C for pyrolysis processes. At low temperature, thermolysis process yields mainly charcoal, which is a valuable energy source. Minor products in the thermolysis process are fuel oil and gases. The chemical equation that represents the thermolysis process of biomass is given below.

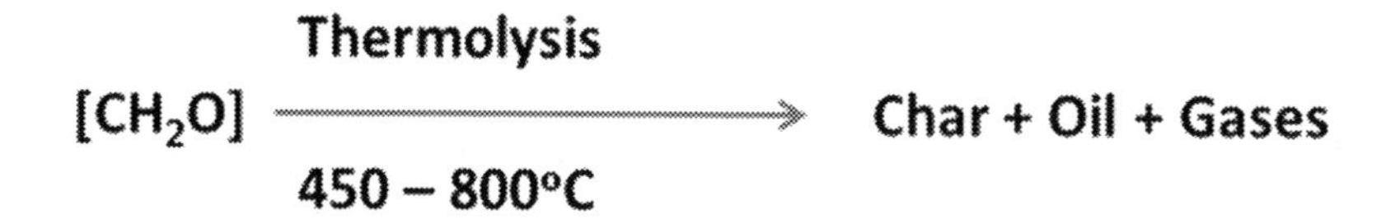

$$[CH_2O] \xrightarrow[450-800^\circ C]{\text{Thermolysis}} \text{Char + Oil + Gases}$$

However, in the thermolysis process 50% of the energy content of biomass is lost. At higher temperature, the pyrolysis process yields gases such as H_2, CO, acetylene as major products with char content as a minor product. The equation representing pyrolysis of biomass is given below.

$$[CH_2O] \xrightarrow[1500^\circ C]{\text{Pyrolysis}} \text{Gases } (C_2H_2) + \text{Char}$$

Gasification of Biomass:

In this process of gasification of biomass involves that conversion of biomass is allowed in air and steam. Thus, the products contain CO and H_2 as major molecules with methane and CO2 as side products. The products find application in electricity generation or chemical feedstock (from synthesis gas). The reaction of gasification of biomass is shown below.

$$[CH_2O] \xrightarrow[\text{Air and Steam}]{\text{Gasification } (660\text{-}1200^\circ C)} CO + H_2 + CO_2 + CH_4$$

Therefore, gasification of biomass is completely different from thermolysis and pyrolysis wherein O2 is completely eliminated.

Hydrothermolysis:

This process was originally developed by Shell oil company to produce an oil-like material (bio-crude). In the hydrothermolysis of biomass needs a low oxygen content and requires high temperature (200 – 330°C) and higher pressure (30 bar) in a closed aqueous medium. Essentially, biomass is subjected to hydrothermal treatment in low oxygen content. Hydrothermolysis of biomass written in an equation is shown below.

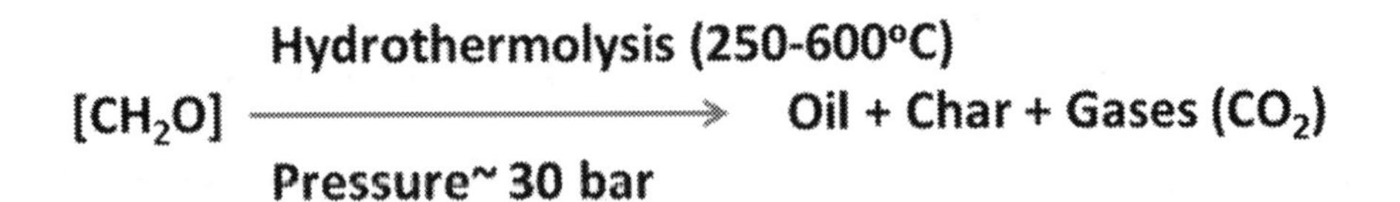

Anaerobic Digestion:

Anaerobic digestion refers to the treatment of biomass with bacteria in the absence of air to produce methane rich gas.

Bioethanol:

It was well-known that fermentation of glucose based crops such as sugar cane and corn starch using Saccharomyces yeasts lead to produce ethanol. This process of involvement of fermentation process to produce ethanol is known as bioethanol. In the USA, dry bioethanol (or anhydrous ethanol) is mixed with gasoline (up to 22%) and this blend is being used in the conventional engines.

Biodiesel:

Rudolf Diesel invented the first engine which ran on groundnut oil (biodiesel). Because of high viscosity problem associated with pure vegetable oil, diesel engine cannot run satisfactorily. Also, it has poor cold-start properties. Triglycerides from vegetable oils constitute major components. Also, on long standing the triglycerides get viscosity increased due to oxidation and polymerization of the molecules. Therefore, raw vegetable oils are converted into useful material by transesterification technology. Ester (fatty acid ester) is separated from vegetable oils by treating it with low molecular weight methanol in the presence of a catalyst. Thus, the ester is used as biodiesel.

Hydrogen:

Hydrogen gas is considered as a forever fuel because of availability of hydrogen in different forms. Therefore, research in America and rest of the world is directed to generate, store and applications of hydrogen gas. To explore hydrogen gas as a fuel, either internal combustion engine or fuel cell is required to get energy from a straight forward reaction between hydrogen and oxygen gases. This reaction yields only water vapor. The ultimate goal in hydrogen economy is to replace currently available dependent and carbon-based fuel by independent hydrogen gas fuel. This replacement is also proved not to produce greenhouse gas such as CO_2.

By 2020, the world will face an oil crisis. The majority of American automobile owners will suffer greatly from this crisis if new energy sources are not available. The first oil shortage in the 1970s and 1980s was economically and politically induced. This time, however, the crisis will be based on a real shortage of oil for fuel. Although optimist argue that new oil fields like exist 3280 feet or more below the surface of the oceans, the process of finding and obtaining it is very expensive and the technologies to do so are not fully developed. Therefore, it is highly desirable to avoid dependence of fossil fuel. Also, carbon-based oil produces carbon dioxide (CO_2) as a byproduct that will likely increase global temperature by 2.52 to 10.44°F over the next one hundred years. These temperature increased will cause the polar ice caps to melt. Therefore, alternative to carbon based oil fuels must be researched. This could be achieved by using Hydrogen (H_2) as a fuel that produces only healthy water vapor when it burns, and does not heat up the air.

Chemical reactions could be explored to produce hydrogen gas. One such example is the violent reaction of reactive metal with acid. Thus, Zn reaction with HCl yields $ZnCl_2$ and H_2 gas. Similarly, Fe metal reacts with sulfuric acid leads to production of H_2 gas and $FeSO_4$. These reactions are represented below.

$$Zn + 2HCl \rightarrow ZnCl_2 + H_2\,(gas)$$
$$Fe + H_2SO_4 \rightarrow FeSO_4 + H_2\,(gas)$$

Another chemical reaction is that metals such as K, Na and Ca react vigorously with water producing enough heat to ignite the hydrogen. Example of Na reaction with water is represented by the following reaction.

$$2Na + 2H_2O \rightarrow 2NaOH + H_2\,(gas)$$

The formed NaOH in the above mentioned reaction further can be treated with Al or Si in presence of water to produce H_2 gas.

e.g. $2Al + 2NaOH + 6H_2O \rightarrow 2NaAl(OH)_4 + 3H_2\,(gas)$

$$2NaOH + Si + H_2O \rightarrow Na_2SiO_3 + 2H_2\,(gas)$$

Hydrides could be explored as a source for hydrogen. Calcium hydride or sodium borohydride is treated with water to produce hydrogen gas and the reactions involved in these processes are represented below.

$$CaH_2 + 2H_2O \rightarrow Ca(OH)_2 + 2H_2\,(gas)$$
$$NaBH_4 + 4H_2O \rightarrow NaB(OH)_4 + 4H_2\,(gas)$$

Hydrogen on either burning with Oxygen gas yields tremendous heat which is ultimately an energy source and electrochemical reaction be between hydrogen at anode and oxygen at cathode in the fuel cells also generates electricity. Fig.5 below shows basic design of a fuel cell.

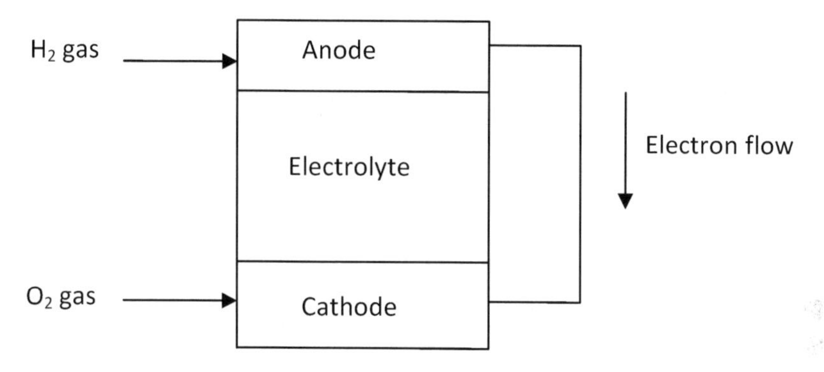

Fig. 5: Schematic diagram of a fuel cell

Anode:

Anode consists of carbon support and on which Platinum catalyst is deposited. Hydrogen entering the anode is converted into hydrogen ions by platinum catalyst and the hydrogen ions pass into the electrolyte layer. The released electrons thus from hydrogen molecules can't enter into electrolyte while it enters through external circuit generating electricity into cathode.

Cathode:

Oxygen gas enters at the cathode where in it combines with protons and electrons to form water. The cathode is made up of metal oxides that are stable in oxygen environment.

Electrolyte:

There are various types of electrolytes employed in the fuel cell and hence, electrolyte type determines the operating temperature of fuel cell and name of fuel cell. Table 5 below summarizes types of fuel cell, electrolyte and operating temperature.

Table 5: Summary of various type (or name) of fuel cell with its electrolyte and operating temperature.

Electrolyte	**Operating temperature, °C**	**Name of fuel cell**
Concentrated H_3PO_4	~200°C	Phosphoric acid fuel cell
Ceramic ZrO_2	~1000°C	Solid oxide fuel cell
Conducting Polymer	80 – 90°C	Proton exchange membrane fuel cell
Ca or Li/K carbonates	>600°C	Molten carbonate fuel cell
25 – 85% KOH	80°C	Alkaline fuel cell

These fuel cells are very briefly outlined one by one.

Phosphoric Acid Fuel Cell:

Phosphoric acid fuel cell generates a few megawatts of electricity and hence, it finds applications in small power generating projects. Examples include applications in hospitals and offices. Highest efficiency of electricity generation can be obtained using 100% phosphoric acid as the electrolyte supported between poly(tetrafluoroethene) (PTFE) plates at above 200°C and little above atmospheric pressure. The anode is Platinum supported on carbon black and the cathode is PTFE. Anode is readily poisoned by low levels of CO. Also, carbon black faces corrosion problem.

Solid Oxide Fuel Cell:

Solid oxide fuel cell operates only above 700°C. The high temperature operation of solid oxide fuel cell has the following advantages.

a. suitable for steam production
b. Ability to reform many fuels efficiently
c. Overall fast kinetics of the system

However, it has a main drawback of choices of materials in use in solid oxide fuel cell.

Here, the solid electrolyte is a ceramic one. Thus, the ceramic electrolyte is composed of zirconia stabilized with Yttria so that high temperature cubic phase is stabilized. The anode consists of physical mixture of zirconia and nickel. The cathode is composed of doped $LaMnO_3$ ceramic.

Proton Exchange Membrane fuel cell:

Proton exchange membrane fuel cell is also called as polymer electrolyte fuel cell or the solid polymer fuel cell. Thus, polymer electrolyte employed in this type fuel cell is perfluorinated sulphonic acid. It has chemical stability and proton conduction capability. Its operation temperature is only about 80°C and hence, platinum loaded or impregnated onto carbon or PTFE has highly efficient electrodes function. Highly pure hydrogen at the anode and highly pure oxygen at the cathode are preferred due to poisoning of other fuels at the anode. A major drawback of the fuel cell is requirement of hydrated polymer electrolyte to transfer proton.

Molten carbonate fuel cell:

Mostly lithium carbonate or potassium carbonate is used as electrolyte. Its operating temperature is around 650°C due to requirement of molten carbonate to provide good conductivity. Nickel alloys as anode and lithium oxides as cathode are explored in the molten carbonate fuel cell. It finds application in commercial power production.

Alkaline Fuel Cell:

Potassium carbonate is used as electrolyte. Therefore, it readily poisons in contact with CO_2 molecules. Therefore, highly pure Oxygen gas at the cathode is essential. Because of its efficiency it finds applications in space and military and even today the space shuttle explores alkaline fuel cell to power it and yields pure drinking water at the cathode side.

Questions:

1. Describe classification of energy sources
2. What is petroleum refining?
3. What is atmospheric distillation?
4. How does atmospheric distillation differ from vacuum distillation?
5. Outline thermal processes in the petroleum refinery industry
6. How does catalytic cracking differ from thermal cracking?

7. Describe three methods of catalytic cracking
8. State Deasphalting & Dewaxing
9. Compare hydro treating with hydrocracking
10. What is reforming? Give an example
11. State catalytic reforming
12. What is calorific value of fuel?
13. Give and elaborate alternate energy sources with an example.

Spectroscopy

Objectives:

1. To introduce spectroscopy
2. To state infrared absorption & its corresponding infrared spectrum
3. To illustrate infrared spectrum using quantum mechanical approach
4. to outline stretching and bending vibrational modes.
5. to illustrate application of force constant and reduced mass in the Infrared Spectra
6. To state number of vibrational modes for linear and non-linear molecules with examples.
7. To outline infrared spectrometer such as dispersive infrared spectrometer and Fourier transform spectrometer
8. To illustrate infrared spectra of alkanes, alkenes and alkynes.
9. To state difference between Raman and Infrared Spectroscopy
10. To outline basic theory of Raman Spectroscopy
11. To define Raman active molecules with a few examples
12. To define principle of LASER action with types of LASER
13. To compare Raman Spectra with Infrared Spectra
14. To state the principle of Nuclear Magnetic Resonance (NMR) Spectroscopy
15. To describe basis of NMR spectroscopy
16. to outline chemical shift and shielding

17. To describe NMR spectrometer
18. To define the meaning of UV-Vis spectroscopy
19. To outline main transitions of electrons for UV-Vis spectroscopy
20. To state Beer-Lambert Law
21. To state the principle of mass spectrometer and mass spectrum.

Spectroscopy of any covalent organic molecules or inorganic solids is dealt with absorption of energy of electromagnetic radiation. Thus, X-ray spectroscopy refers to energy transitions due to bond breaking. Ultraviolet-visible spectroscopy deals with energy transitions due to electronic transition. Infrared spectroscopy treats vibrational energy transition. Microwave spectroscopy invokes rotational energy transitions and Nuclear Magnetic Resonance spectroscopy and Electron Magnetic Resonance Spectroscopy are associated with nuclear spin and electronic spin transitions respectively. This chapter, Spectroscopy thus focuses on the above mentioned various spectroscopies.

Infrared Spectroscopy:

Infrared Absorption:

Molecules are excited from ground state vibrational energy level into higher vibrational energy level due to exposure to infrared radiation of electromagnetic spectrum. This transition takes place within first electronic energy level. The main reason for molecules or compound to absorb infrared region of electromagnetic spectrum is due to vibrational modes of bond between two atoms. This vibrational energy falls in the region of 8 to 40 KJ/moles. Therefore, only molecules can exhibit infrared spectra but infrared spectra can't be observed from the atoms. Thus, molecules are responsible for infrared spectra. The, important question is that what type of bonds are responsible for infrared absorption in the molecules. The answer is simple to state that bonds with dipole moment are capable of absorption infrared energy and thus, molecules such as H_2 or Cl_2 not having dipole moment in the bonds between hydrogen – hydrogen and chlorine – chlorine do not able to absorb energy in the infrared region.

Infrared Spectrum:

Infrared spectrum of a molecule or compound refers to frequency of absorption on X-axis and intensity of the absorption frequency on Y-axis. Since molecules have different stretching and bending vibrational modes, a single molecule can exhibit a particular pattern of absorptions in the infrared region of electromagnetic spectrum. Thus, no molecules or compounds of different structure have the same infrared spectrum.

Quantum Mechanical Approach:

Consider vibrational frequency of a bond in a molecule or compound as a simple harmonic oscillator. Thus, vibrational energy levels of simple harmonic oscillator are equally spaced. The mathematical expression for energy levels of simple harmonic oscillator is given as

$E_n = (v+1/2)h\upsilon$, where v = 0, 1, 2, 3, ……

Where h = Planck's constant

υ= Vibrational frequency

v= quantum number

when v= 0, vibrational energy level, E_0 is equal to ½ $h\upsilon_0$. Interestingly, vibrational energy is not equal to zero even at the lowest vibrational energy level. Therefore, the lowest vibrational energy is called zero point energy.

However, in the real system, non-harmonic oscillator is observed. Therefore, the energy levels of non-harmonic oscillator will be different from the energy levels of a harmonic oscillator. Thus, with increasing energy, the energy levels of non-harmonic oscillator are compressed.

Vibrational stretching and bending modes:

Usually, stretching vibrational mode and bending vibrational mode are responsible for infrared absorption or infrared active in the electromagnetic spectrum. Figure below represents stretching vibration of C – H bond and bending vibration of C – O –H bond angle.

Stretching vibration can be further divided into symmetric stretching and asymmetric stretching vibration. Thus, among asymmetric, symmetric and bending vibrational modes, asymmetric stretching falls in the highest energy followed by symmetric stretching then bending vibrational mode energy follows.

Vibrational Absorption trends:

Consider a bond between two atoms. Vibrational absorption energy of this bond depends upon the bond strength and reduced mass of the two atoms.

The total amount of vibrational energy is proportional to the frequency of the vibration as shown below.

$$E_{vib} \propto h\nu_{vib}$$

For harmonic oscillator, the vibrational frequency is given as

$$\nu_{vib} = \frac{1}{2\pi C} \frac{K}{\mu}$$

Where K = force constant of the bond and

μ = reduced mass of the two atoms

According to Hooke's law, the reduced mass, μ is given by

$$\mu = \frac{m_1 m_2}{m_1 + m_2}$$

Where m1 = mass of atom 1

m2 = mass of atom 2.

Thus, vibrational energy is directly related to square root of force constant and is inversely related to square root of reduced mass of the two atoms.

The application of force constant and reduced mass in comparison of frequency of vibration is explained one by one with examples.

Application of force constant:

The force constant refers to the strength of the bond between two atoms. As number of bonds between the same two atoms increases, the force constant increases. A typical example is that the force constant, K increases with increase in number of bonds between carbon – carbon atoms as shown below.

C — C < C ═ C < C ≡ C

⟶

Force Constant, k increases

Applying this force constant increment with increase in number of bonds in the equations of E_{vib} and ν_{vib}, the energy of frequency of vibration increases with increase in number of bonds between carbon and carbon atoms. Therefore, in reality, absorption of vibrational frequency of triple bonds between carbon and carbon is 2150 cm^{-1}, of double bonds between carbon and carbon is on is 1650 cm^{-1} and of single bond between carbon and carbon is 1200 cm^{-1}. Bending frequency occurs at lower vibrational energy than stretching frequency due to lower force constant, k for bending frequency. For example, H - C – H stretching is observed at 3000 cm^{-1} whereas its bending is observed at 1340 cm^{-1}.

Hybridization of carbon atom affects the force constant and thus, C – H vibration increases with decrease in hybridization as shown below.

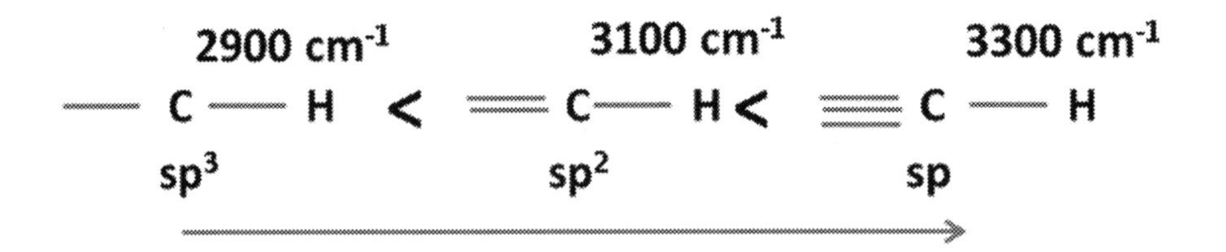

Force Constant, k increases

Application of reduced mass:

Consider reduced mass of two atoms that increase as shown below.

$$\mathrm{C-H} < \mathrm{C-C} < \mathrm{C-O} < \mathrm{C-Cl} < \mathrm{C-Br} < \mathrm{C-I}$$

Reduced mass, μ increases →

Applying this reduced mass increment in the equations of E_{vib} and ν_{vib}, the energy of frequency of vibration decreases with increase in reduced mass. Therefore, in reality, absorption of vibrational frequency of C – H is at 3000 cm^{-1}, of C – C is at 1200 cm^{-1}, of C – O is at 1100 cm^{-1}, of C – Cl is at 750 cm^{-1}, of C – Br is at 600 cm^{-1} and of C – I is at 500 cm^{-1}.

Normal modes of Vibration:

Normal modes of vibration refer to the set of fundamental vibrations, which are based on three set of coordinate axis. Thus, normal modes of vibration of any molecule can be defined as shown below.

For non-linear molecules,

Number of normal modes of vibration = 3N – 6

For linear molecules,

Number of normal modes of vibration = 3N – 5

Where N = number of atoms in the molecule.

Examples:

(a) H_2O, a non-linear molecule:

Water, H_2O molecule is an example for a non-linear molecule as shown below.

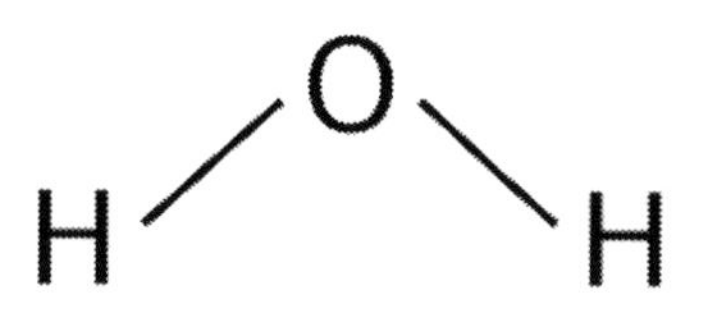

Number of atoms in the water molecule, H_2O = N = 3.

Therefore, number of normal modes of vibration = 3N – 6 = (3 x 3) – 6 = 9 – 6 = 3.

Thus, number of normal modes of vibration for water molecule = 3.

(b) CO_2, a linear Molecule:

CO_2 molecule is an example for a linear molecule as shown below.

O=C=O

Number of atoms in the Carbon dioxide molecule = N = 3

Therefore, number of normal modes of vibration = 3N – 5 = (3 x 3) – 5 = 9 – 5 = 4.

Thus, number of normal modes of vibration for carbon dioxide molecule = 4.

Infrared Spectrometer:

The instrument that is used to determine absorption spectrum of vibrational moles is called infrared spectrometer. There are two types of infrared spectrometer known and these are

a. Dispersive Instrument
b. Fourier Transform Instrument.

Both the instrument can provide vibrational absorption spectrum between 4000 – 400 cm-1. But, Fourier transform instrument is a recent instrument with more rapid recording spectrum.

a. Dispersive Infrared Spectrometer:

The dispersive infrared spectrometer consists of the following components.

(i) Infrared Source
(ii) Infrared radiation divider
(iii) Monochromator
(iv) Detector and
(v) amplifier and Recorder

(i) Infrared Source:

A hot wire is a source for infrared radiation. It produces a beam of infrared radiations.

(ii) Infrared radiation divider:

Infrared radiation divider consists of mirror. The mirror divides the infrared radiation into two parallel beams with equal intensity. One beam is passed through sample and another beam is passed through reference.

(iii) Monochromator:

Monochromator consists of a rapidly rotating sector (chopper), which allows alternately radiation from sample and reference to the diffraction grating.

(iv) Detector:

Detector for infrared radiation is thermocouple. The detector can able to differentiate the ratio of intensities between reference and sample beams. Thus, detector can detect which frequencies have been absorbed by the sample and the frequencies are not affected by the sample.

(v) Amplifier and Recorder:

The signal from detector is sent to amplifier and finally, it leads to resulting spectrum of the sample in a chart.

Figure 1 below outlines a schematic diagram of dispersive infrared spectrometer.

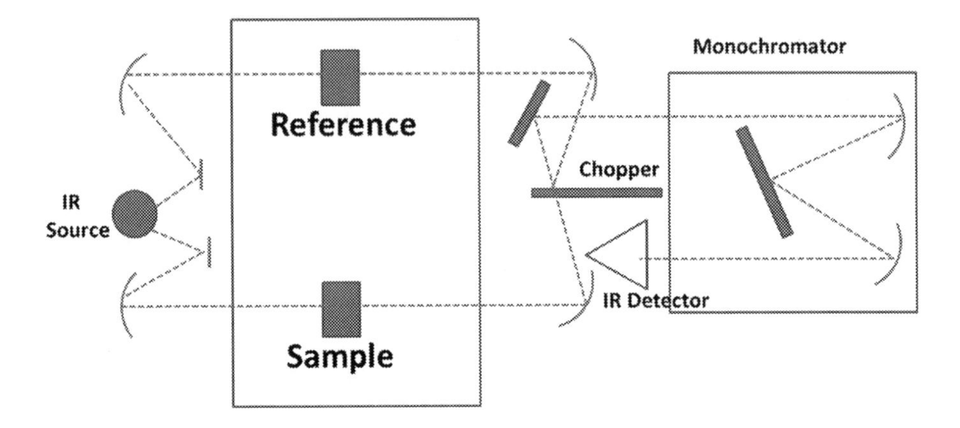

Fig. 1: Outline of Dispersive Infrared Spectrometer.

(b) Fourier Transform Spectrometer:

Fourier Transform (FT) Infrared consists of interferogram. Interferogram refers to intensity on y-axis and time on x-axis. This type of spectrum is called time-domain spectrum. But, spectrum containing intensity on y-axis and frequency on x-axis is required for chemists and this type of spectrum that consists of intensity – frequency is called frequency- domain spectrum. Fourier transform is a mathematical operation which converts time – domain spectrum into frequency – domain spectrum and the infrared spectrometer that explores Fourier transform operation is called FT – IR spectrometer. Merits of FT – IR over dispersive infrared spectrometer are

1. FT - IR spectroscopy is a rapid technique and
2. FT – IR has greater sensitivity.

Infrared Spectra of alkanes, alkenes and alkynes:

In order to identify the hydrocarbon of alkanes, alkenes and alkynes, C – H stretching and bending frequencies of absorption are important to monitor in the infrared spectrum. As known from hydrocarbon chemistry that C – H stretching of sp^3 carbon of alkane has absorption of stretching vibration of 2900 cm^{-1}, C – H stretching of sp^2 carbon of alkene has the absorption of stretching vibration of 3000 cm^{-1} and C – H stretching of sp carbon of alkyne has absorption of stretching vibration of 3300 cm^{-1}. Table below summarizes type of carbon involved in the C – H stretching, its corresponding hybridization of carbon, bond length, its strength and infrared frequency of absorption.

Table 1: Summary of C – H stretching vibrations of sp^3, sp^2 and sp carbon.

Type of hydrocarbon	Type of carbon	Bond length	Bond Strength	IR frequency of absorption
Alkane	Sp^3	1.12 Å	422 KJ	2900 cm^{-1}
Alkene	Sp^2	1.10 Å	444 KJ	3100 cm^{-1}
Alkyne	Sp	1.08 Å	506 KJ	3300 cm-1

Vinyl carbon, aromatic carbon and cyclopropyl carbon have all sp^2 carbon present in them as shown their structures below.

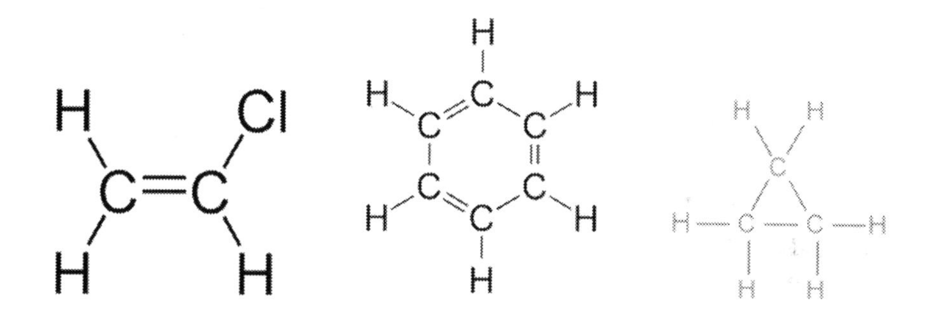

Vinyl Carbon with C – H bond Aromatic carbon with C – H bond Cyclic carbon with C – H bond

C – H stretching vibration of the sp^2 carbon is different from each other. Thus, infrared absorption spectrometer can be used to identify the type of C – H stretching from these sp^2 type carbons. Thus, increase in strain from vinyl =C – H to aromatic = C – H and then to cyclopropyl C – H shifts the stretching vibration frequency from 3100 cm^{-1} towards 3300 cm^{-1}.

Stretching vibrations for various sp^3 hybridized C – H bonds:

Methyl (CH_3), methylene ($-CH_2$) and methine ($(CH_3)_3C$ - H) represents various sp3 hybridized C – H bonds. Methine C – H stretching vibration is usually observed at 2890 cm^{-1}. When methylene C – H stretching vibration is considered, two vibrational stretching frequencies are observed and these are at 2926 cm^{-1} due to asymmetric stretching and 2853 cm^{-1} due to symmetric stretching. Thus, methine C – H stretching vibration splits into 2926 cm^{-1} (value is higher than methine C – H stretching vibration) and 2853 cm^{-1} (lower than methine C – H stretching vibration).

When methyl C – H vibration stretching frequencies are considered, again two vibrational bands are observed. One is at 2962 cm^{-1} due to asymmetric stretching and another is at 2972 cm^{-1} due to symmetric stretching vibration. Thus, methane C – H stretching vibration splits into 2962 cm^{-1} (higher than methine C – H stretching vibration) and 2872 cm^{-1} (lower than methine C – H stretching vibration).

In the case of methylene and methyl C – H vibration stretching frequencies, asymmetric stretching generates larger dipole moment and hence, intensity of absorption of asymmetric stretching vibration is greater.

In addition to C – H stretching vibrations of methyl and methylene groups, bending vibration of methyl and methylene groups can be explored to identify them. Thus, CH_2 scissoring bending shows vibrational frequency absorption at 1465 cm^{-1} and one of the bending of methyl group, CH_3 shows vibrational frequency absorption at 1375 cm^{-1}.

C = C stretching Vibrations:

Effect of alkyl substituted alkenes:

C = C stretching vibration depends upon other two bonds on each sp2 carbon. Thus, C = C stretching of non-substituted alkenes have absorption values near 1640 cm-1. 1,1 – disubstituted alkenes have C = C stretching vibration at about 1650 cm-1. But, tri and tetra substituted alkenes have C = C stretching vibration values near 1670 cm-1. Thus, as substitution increases, C = C stretching vibration values increase. But, intensity of C = C stretching vibration decreases as substitution on alkenes increases and the intensity, in fact, of C = C stretching vibration is lower than that of C = O bon stretching vibration.

Effect of Cis and Trans isomers of alkenes:

Between cis and trans isomers, no change in dipole moment occurs with cis isomer during its stretching and hence, there is no infrared absorption observed for cis isomer. But, it is not the case with trans isomer due to observation of stretching frequency of absorption.

Effect of conjugation of alkenes:

When conjugation is introduced to alkenes by C = C bond or C = O bond, force constant, k of C = C bond is lower and hence, conjugated double bonds have lower frequency of vibration.

Effect of C = C double bon in cyclic compounds:

The effect of double bonds on vibration frequency of absorption of acyclic compounds was discussed. When cyclic compounds are considered, double bond presents in the different ring sizes affect the C = C vibrational frequency of absorption. It is noted that double bonds within cyclic compounds are called endo double bonds. Thus, it is interesting to note that as internal angle of ring decreases, the vibrational frequency of absorption decreases. This trend is true until the internal angle reaches down to 90°. Figure below summarizes endo double bonds effect on vibrational frequency of absorption in size of cyclic compounds.

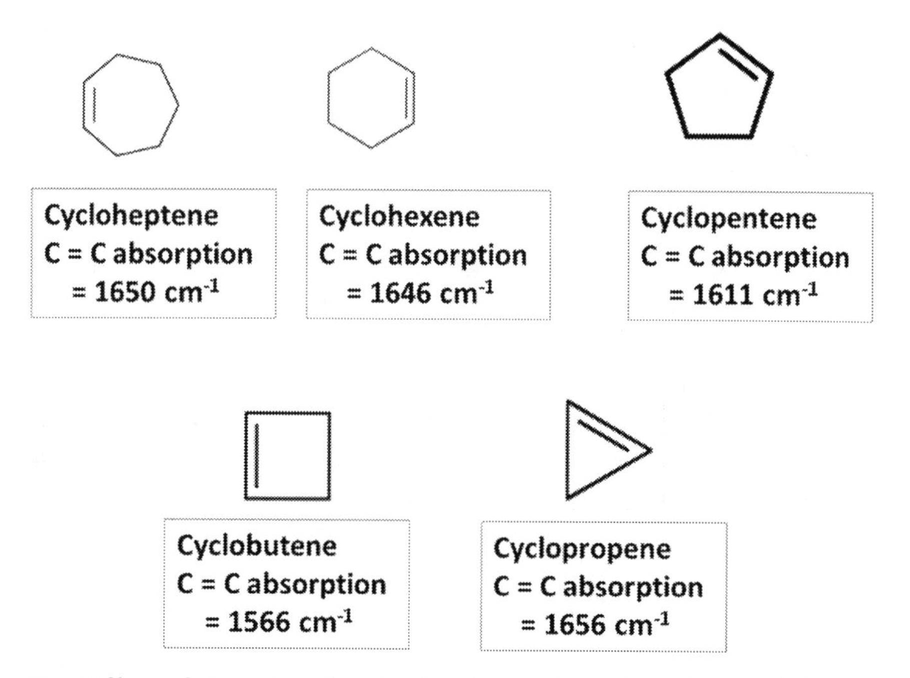

Fig.: Effect of ring size of endo double bonds on its vibrational frequency absorption

C = C bond in cyclopropene has exceptionally highest vibrational frequency of absorption.

The vibrational frequency of exo C = C double bonds increases with decreasing ring size.

Table 2 below summarizes vibrational frequency of absorption of some functional groups.

Table 2: Vibrational frequency of absorption of selected functional groups.

Functional Group	**Vibrational Frequency of absorption, cm^{-1}**
OH	3400 – 3700
NH_2	3200 – 3500
CH	2800 – 3100
C = N	1600 – 1700
C = O	1650 – 1750
C = C	1550 – 1750
Aromatic C = C	1450 – 1600
C – O	1000 - 1300

Raman Spectroscopy:

Absorptions due to vibration of molecules are determined based on the two spectroscopies such as infrared spectroscopy and Raman spectroscopy. As seen in the earlier chapter, infrared spectroscopy involves absorption of infrared radiation due to excitation of molecules from lower vibrational level to higher vibrational level in the ground state electronic levels. In the case of Raman spectroscopy, scattering of visible radiation (light) is determined by molecules but in the visible region. Thus, Raman scattering is due to inelastic scattering of molecules which is in contrast to Rayleigh scattering (Elastic scattering) of molecules.

Basic Theory:

When light interacts with matter, light is either absorbed by the matter or scattered by the matter or passed through the matter, where light refers to visible radiations of electromagnetic spectrum only. The reason for light absorption by the matter is that the incident light energy is equal to energy gap between ground and excited electronic states. The absorption of light is usually measured by absorption spectroscopy. Light scattering by matter does not to be matching with energy gap of molecules. Thus, energy difference between incident radiation of light and scattered radiation of light is referred to Raman Shift, which is in the region of vibrational energy of molecules.

Rayleigh scattering is due to distortion of electrons cloud whereas Raman scattering is due to distortion of nuclei cloud. Therefore, Rayleigh scattering does not alter the energy of incident radiation of light (Raman Shift = 0) and Raman scattering does change the energy of incident radiation of light (Raman Shift is not equal to zero). The change of energy of incident light or Raman shift is equal to vibrational energy of molecules. If Raman shift is positive, it is called Stokes shift (or lines). If Raman shift is negative, it is called Anti Stokes Shift (or lines). If Raman shift is zero, it is called Rayleigh scattering. Thus, Raman scattering refers to Stokes and Anti Stokes lines.

Interestingly, the energy difference between stokes lines and Rayleigh scattering or the energy difference between Anti Stokes lines and Rayleigh scattering is equal to vibrational energy of molecules. The most importantly, thus, vibrational energy of molecules in the Raman spectroscopy is observed in the visible region. Raman scattering (Stokes lines and Anti Stokes lines) and Rayleigh scattering are explained briefly using a diagram below.

At room temperature, mostly, molecules occupy at the lowest vibrational levels. When the light interacts with molecules, ***virtual energy levels are created temporarily*** and thus, molecules are excited from the ground vibrational level to virtual energy levels by incident radiation. When the molecules return to higher vibrational level Stokes lines are formed. Similarly, molecules are excited from the higher vibrational level to virtual energy levels by incident radiation. When the molecules return to lower vibrational level Anti Stokes lines are formed. Since the probability of molecules at the lower vibrational level is higher than that of molecules at the higher vibrational level and hence, usually, Stokes lines are stronger and intense than that of Anti Stokes lines. In the case of Rayleigh scattering the molecules are excited from the ground vibrational level to virtual energy levels by the incident radiation and the molecules returns to the same ground vibrational level. Therefore, in principle and in practice, intensity of Rayleigh scattering is higher than that of Raman Scattering (Stokes and Anti Stokes lines). Figure 2 below illustrates the Rayleigh scattering and Raman scattering.

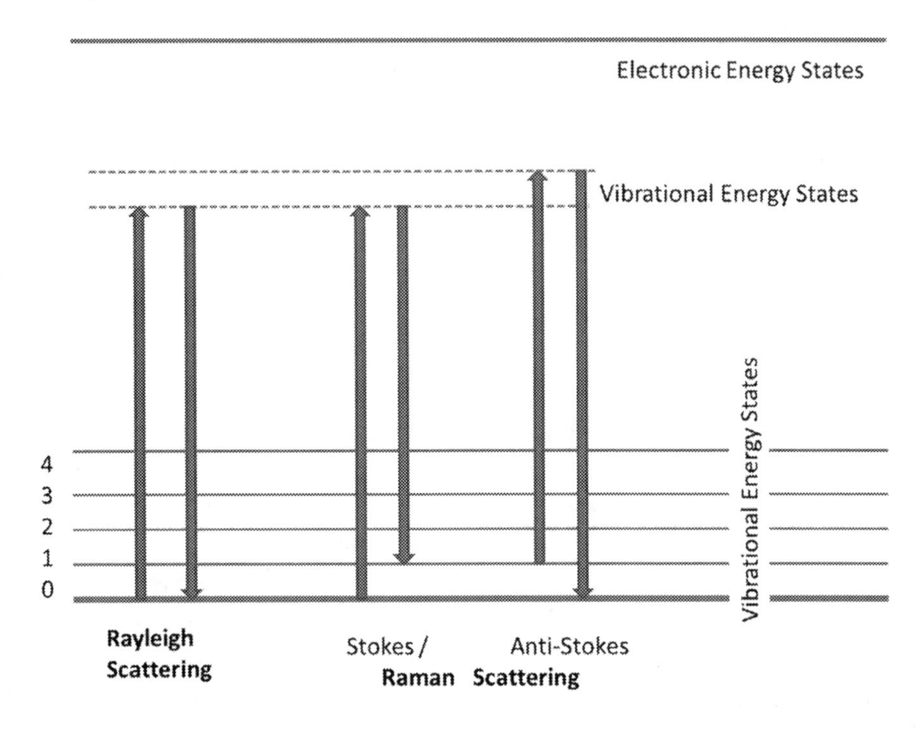

Fig.2: Difference between Raman scattering and Rayleigh scattering

Raman Active Molecules:

It is straight forward that molecules with dipole moment or change in dipole moment during vibration are infrared active. But, molecules to be active in Raman spectrum should show molecules with change in polarization during stretching. Thus, polarization change of molecules is considered in Raman spectrum and change in dipole moment is required to be active for molecules in infrared spectrum. The selection rule is that

Vibration should change polarization of molecules to be active in Raman Spectrum

Vibration should change dipole moment of molecules to be active in Infrared spectrum.

Example One:

Consider O_2 (oxygen) molecule. O_2 molecule does create polarization change during stretching vibration and hence, O_2 molecule is active in Raman Spectrum. But, O_2 molecule does not have dipole moment or does not change dipole moment during stretching vibration and hence, O_2 molecule is not active in Infrared spectrum.

Thus, homonuclear diatomic molecules are active in Raman spectrum but they are not active in infrared spectrum.

Example Two:

Consider linear CO_2 (carbon di oxide) molecule. Stretching vibration of CO_2 molecules makes polarization change due to nuclear displacement of molecules. Therefore, stretching vibration of CO_2 molecule shows strong peak in Raman Spectrum. Bending vibration of CO_2 molecule create large dipole moment of CO_2 molecules but little change in polarization of the CO_2 molecule. Therefore, bending vibration of CO_2 molecule shows strong peak in Infrared spectrum but little intense or no peak is observed in Raman spectrum for bending vibration of CO_2 molecule.

LASER Action:

With the advent of LASER, the Raman Spectroscopy became in reality. Thus, the highest monochromatic light from LASER is the source for incident radiation for the Raman Scattering. Thus, Raman Spectrometer contains the LASER source of light and scattered light is analyzed at perpendicular to the incident light to get Raman Scattering.

Principle of LASER Action:

LASER refers to Light Amplification by the Stimulated Emission of Radiation. Thus, stimulated emission of radiation leads to amplification of electromagnetic radiation of light. The basis for this observation of amplification of electromagnetic radiation is due to absorption and emission of radiation. When the incident radiation energy is resonant to ground state of a system absorption of incident radiation takes place. From the excited state, system spontaneously returns to the ground with with spontaneous emission. The characteristic of spontaneous emission of radiation is that the radiation is emitted by a wide region of solid angle and the radiation has none of directional properties of Laser characteristics. In addition spontaneous emission, the stimulated emission of radiation is possible. If excited energy levels are due to photon absorption, electron bombardment etc. then, these excited states interact with resonant of photon and the system thus returns to the ground state and thereby, amplification of incident radiation takes place in the stimulated emission.

Types of Laser:

Solid state Lasers:

Cr^{3+} doped Al_2O_3 (Ruby) was the first solid state laser. But, today, the most commonly explored solid state laser is Nd^{3+} doped Yttrium Aluminum Garnet Oxide (YAG:Nd^{3+} laser). These solid lasers usually have broad absorption and hence, it is possible to pump these solid lasers by optical means of Flash lamp.

Gas Lasers:

Gas lasers usually contain lines. Because of sharp absorption of gaseous molecules, electrical methods are explored to pump the gas lasers. A typical example for gas lasers is He/Ne or I laser.

Dye Lasers:

In the dye lasers, fluorescent material dissolved in solvent acts as dye laser. Using several chemical method of modification of dye molecules, dye lasers can be tuned to our convenient and hence, among the several types of lasers, dye lasers are preferred one.

Comparison between Raman and Infrared spectra:

Table 3 below compares between Raman and Infrared spectra.

Table 3: Comparison between Raman Spectra and Infrared Spectra

Raman Spectra	**Infrared Spectra**
Raman Spectra are due to scattering of visible light	Infrared spectra are due to absorption of infrared
Raman spectra provide frequency of vibration in the visible region	Infrared spectra provide frequency of vibration in the Infrared region
In the Raman spectra, the frequency of vibration is equal to Raman Shift	In the Infrared spectra, the frequency of vibration is equal to absorption of infrared

Nuclear Magnetic Resonance Spectroscopy:

Nuclear magnetic resonance is yet another spectroscopic method that is widely used now-a-days by organic chemists in addition to infrared spectroscopy. Magnetically different nuclei can be distinguished by nuclear magnetic resonance spectroscopy.

Like electron spin, atomic nucleus with odd mass or odd atomic number or both odd mass and odd atomic number has quantized spin angular momentum and a magnetic moment. Typical examples of such atomic nucleus include ${}^{1}_{1}H$, ${}^{13}_{6}C$, ${}^{14}_{7}N$, ${}^{17}_{8}O$ and so on. It is noted that common isotope nuclei such as ${}^{12}_{6}C$ and ${}^{16}_{8}O$ do not have either odd mass or odd atomic number and hence, ${}^{12}_{6}C$ and ${}^{16}_{8}O$ are not considered in the nuclear magnetic resonance spectrum.

Nuclear spin of nucleus is responsible for quantized spin angular momentum. The number of quantized spin states is determined by nuclear spin quantum number, I. From the nuclear spin quantum number, I, the number of spin states are calculated using 2I+1 with integral spin differences ranging from –J to +J.

Example One:

Calculation of number of spin state of ${}^{1}H$:

Proton has one spin and hence, nuclear spin quantum number, I = ½. The number of spin states for proton is equal to

$$2I+1 = 2(1/2)+1 = 2.$$

Therefore, the spin states for ${}^{1}H$ are +1/2 and -1/2.

Example Two:

Calculation of number of spin states for $^{35}{}_{17}Cl$:

Chlorine with isotope of $^{35}{}_{17}Cl$ has nuclear spin quantum number, I = 3/2. The number of spin states for this chlorine isotope is equal to

$$2I+1 = 2(3/2)+1 = 4.$$

Therefore, the spin states for $^{35}{}_{17}Cl$ are

3/2, 3/2-1, 3/2-2 and 3/3-3

Further simplifying, the number of spin states for $^{35}{}_{17}Cl$ are 3/2, ½, -1/2 and -3/2.

Table 4 below summarizes the common isotope of nucleus that is being explored in the Nuclear Resonance Spectrum

Table 4: Summary of number of spin states of some common NMR active Nuclei

Isotope	Nuclear Spin Quantum Number, I	Number of spin states, (2I+1)	Allowed spin states, +j, +j-1, --- -J
^{1}H	½	2	+1/2, -1/2
^{13}C	½	2	+1/2, -1/2
^{17}O	5/2	6	+5/2, +3/2, +1/2, -1/2, -3/2, -5/2
^{31}P	½	2	+1/2, -1/2
^{15}N	1	3	+1,), -1

Effect of Magnetic Field on Nuclear Spin states:

The charged particle of nuclei has a magnetic moment, μ due to its charge and spin. For example, ^{1}H has two spin states and they have clockwise (+1/2) and anticlockwise (-1/2) spin and hence, the nuclear magnetic moment of the clockwise spin and anticlockwise spin are not the same. On applying magnetic field, degeneracy of nuclear spin states is lost. One of the spin states is aligned with magnetic field and other of the spin states is opposed to the magnetic field as illustrated below in the figure 3.

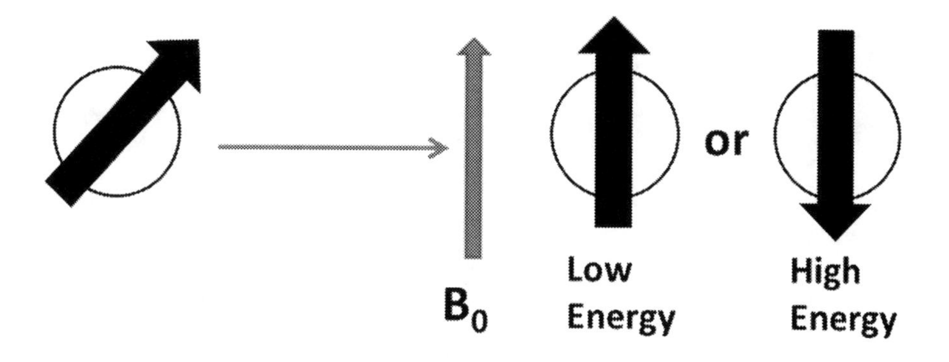

Fig.3:

Therefore, on the application of magnetic field, one nuclear spin state of hydrogen atom becomes the ground state with lower in energy and another nuclear spin state of hydrogen atom becomes the excited state with higher in energy as illustrated below in the figure 4.

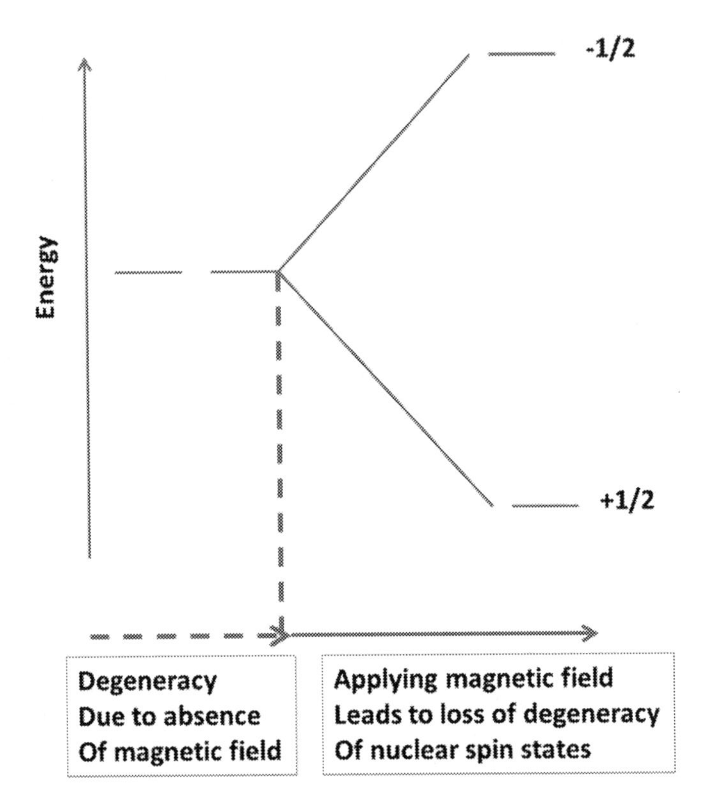

Fig. 4: Effect of applying magnetic field on the nuclear spin states.

Note:

It is noted that optical energy levels such as electronic, vibrational and rotational energy levels are fixed and magnetic energy levels such as nuclear spin states can be altered by applying magnetic field.

Another example to explain the effect of magnetic field on the nuclear spin states of ^{17}O is explained now. ^{17}O isotope of oxygen has six nuclear spin states. All the six nuclear spin states are degenerate with equal energy in the absence of magnetic field. But, on applying magnetic field, the degeneracy of six nuclear spin states is lost and hence, six different spin states are observed on the application of magnetic field as shown below in the Figure 5.

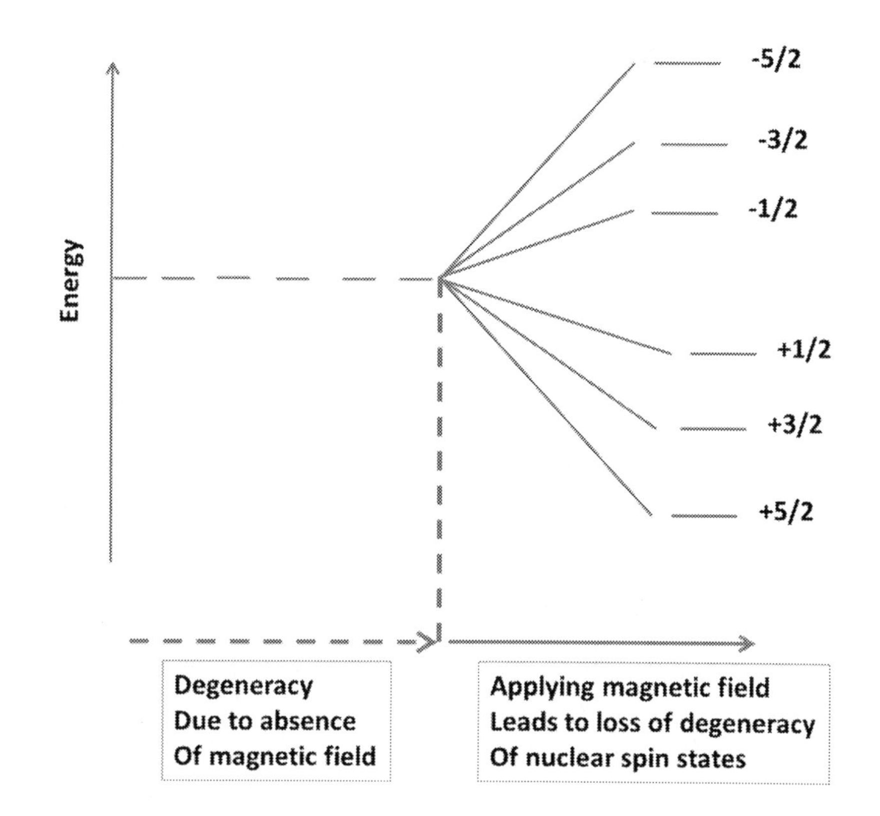

Fig.5: Effect of magnetic field on the nuclear spin states of ^{17}O isotope of oxygen element

Effect of strength of magnetic field on energy of nuclear spin states:

It is established that application of magnetic field leads to loss of degeneracy of nuclear spin states of certain isotope of nuclei. Not, it is highly desirable to study the effect of strength of magnetic field on energy of nuclear spin states. To demonstrate this effect, 1H is considered. Proton has two nuclear spin states with equal energy in the absence of magnetic field. However, as increasing the strength of magnetic field, the energy difference between two nuclear spin states increases as shown below (Fig.6).

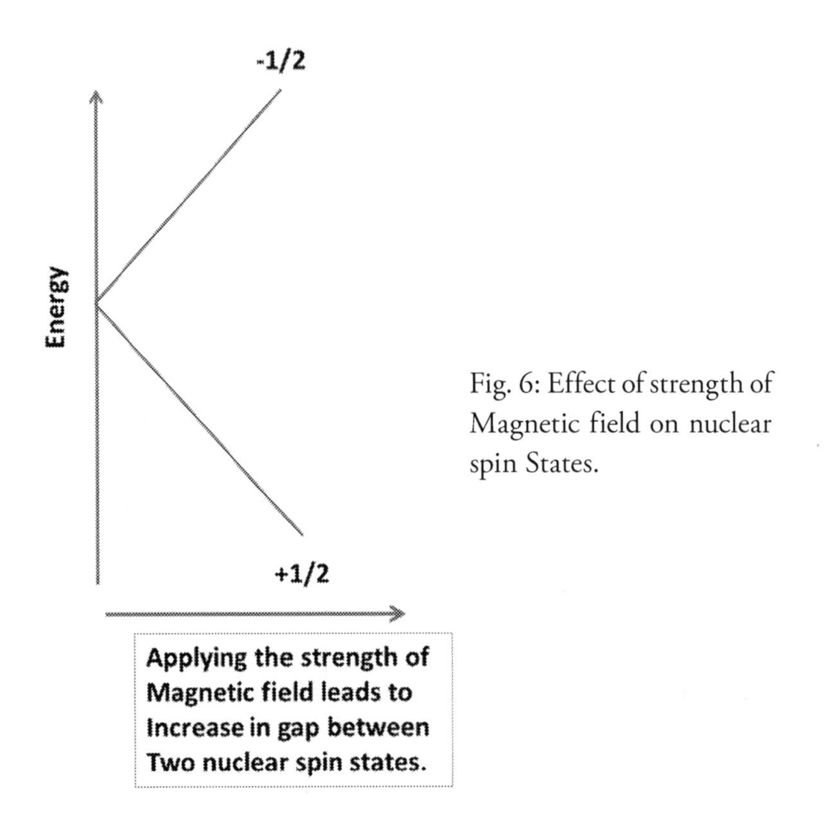

Fig. 6: Effect of strength of Magnetic field on nuclear spin States.

This observation of magnetic strength dependency of energy levels plays an important role in the nuclear magnetic resonance (NMR). Nuclear spin energy separating due to application of magnetic field falls in the Radio Waves. Thus, there are two possibilities to get nuclear magnetic resonance and these are

a. Resonance with change in radio frequency at fixed magnetic field
b. Resonance with change in magnetic field at fixed radio frequency.

Where, resonance refers to absorption of radio waves by nucleus (due to nuclear spin transition from the lower energy to higher energy). The nuclear spin transition depends upon the population densities of nuclear spin states. Usually, nuclear spin state with lower in energy has excess population densities before resonance/absorption of radio frequency. During absorption of radio frequency at a fixed magnetic field, the excess nuclei at the lower nuclear spin state absorbs the radio frequency to go to excited states until the equilibrium or saturation in population densities is reached.

As increasing the radio frequency, the energy difference between two nuclear spin states of ^1H increases and hence, number of excess nuclei also increases as tabulated below (Table 5).

Table 5: Effect of Radio Frequency on the number of excess nuclei

Frequency of Radio Wave	Excess Nuclei
20 MHz	3
60 MHz	9
100 MHz	16
600 MHz	96

The chemical shift and shielding:

All the protons present in a given molecule (especially hydrocarbon molecules) do not have same electronic environment. Thus, valence shell electron densities differ from one proton to another proton present in the same molecule. The valence shell electron densities of each proton generates magnetic field due to their circulation by the application of magnetic field. This magnetic field generated by valence electrons opposing the applied magnetic field is called diamagnetic shielding. The strength of magnetic field of each proton in a molecule is different in the applied magnetic field and hence, the position of nuclear magnetic resonance spectra (NMR spectra) each proton in a molecule is different and this difference is called chemical shift,

The Nuclear Magnetic Resonance Spectrometer:

There are two types of NMR spectrometers and these are continuous wave instrument and pulsed Fourier transform (FT) instrument. The general features of them are described one another one below.

(a) The continuous Wave Instrument:

The solution NMR is a common in use using the continuous wave instrument. Thus, the solvent chosen should be (1) dissolves the compound in interest and (2) does not interfere with the compound in interest. A typical example is Carbon Tetra Chloride (CCl_4) and the internal reference compound used for this purpose is TetraMethylSilane (TMS). There are three basic units for the continuous wave instrument. One is radio frequency generator, second is magnetic field creation and third is radio frequency detector. The solution of the compound with TMS is taken in a small cylindrical glass tube and the glass tube is suspended between the magnetic fields (second component) In order to have same magnetic field experienced by the compound, the glass tube is usually spun around its axis. In the continuous wave instrument, fixed frequency of radio wave (first component) is generator and the radio frequency wave is coiled around the glass tube in the magnetic field. By varying magnetic field, the absorption of radio wave frequency is monitored by the radio frequency detector (third component) that is kept perpendicular to the radio frequency generator. If there is absorption of radio frequency, resonance takes place and hence, the resonance signal is recorded as peak by recorder. Figure 7 below outlines the Nuclear Magnetic Resonance Spectrometer along with typical Nuclear Magnetic Resonance Spectrum of compound, methanol.

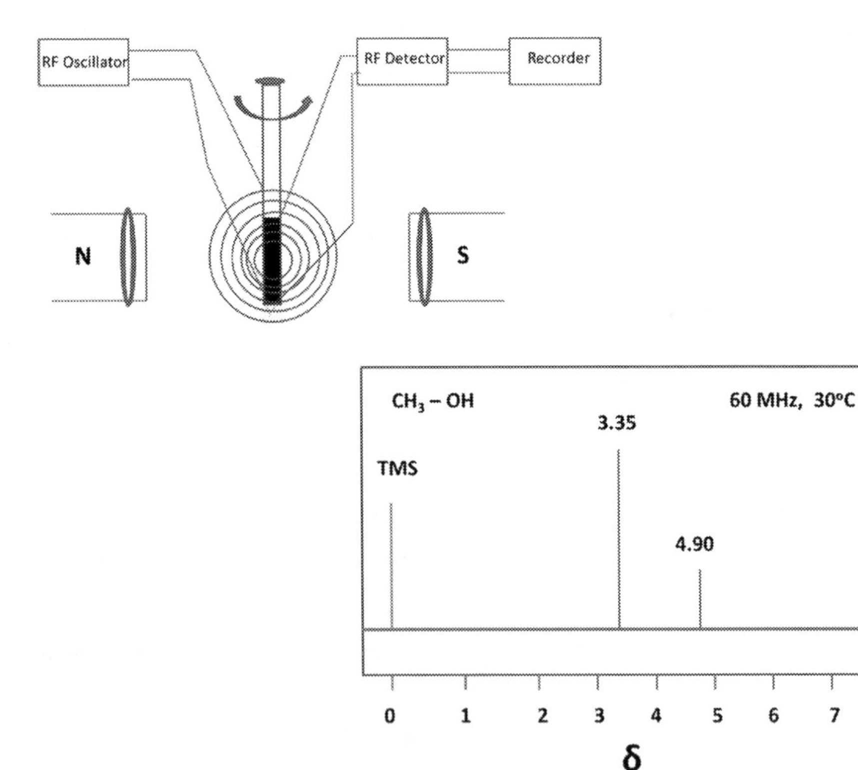

Fig. 7: NMR Spectrometer (left) and NMR spectrum of methanol (right)

Highly shielded protons fall in the right of chart (upfield or high field) while less shielded protons appear in left of the chart (lowfield or downfield).

Pulsed Fourier Transform (FT) instrument:

The FT Instrument of NMR spectrometer is the advenced version of continous wave instrument and it is ofcourse sophisticated instrument. In this one, instread continously varying the magnetic fied from lower strength to higher strength, pulse of energy of magnetic field is provided to excite all the protons simultaneous and thus, resulted in NMR spectra. This sophisticated instrument minimizes the time required to record spectra.

Chemical Equivalence:

All the protons with chemically Identical/equivalence environment exhibit at the same chemical shift or at the same position of NMR spectrum. For example, all the protons present in the benzene molecule and all the protons present in the cylcopentane molecule have identical chemical environment. Therefore, only one peak for each molecule is observed in the NMR spectrum and hence, these protons are called chemical equivence or identical protons.

Magnetic Equivalence:

When sets of protons in a molecule differ from other chemically should have NMR spectrum in a different position are called magnetically equivalence. Sometime it is observed that chemically equivalent protons do not have magnetically equivalence.

Table 6 below compares between optical spectra and Nuclear Magnetic Resonance (NMR) spectra.

Table 6: Comparison between optical spectra and NMR spectra

Optical Spectra	Nuclear Magnetic Resonance Spectra
Energy levels are fixed for gaseous molecules	Energy levels can be altered
Magnetic field is not required to create/ separate energy levels	Magnetic field is required to separate the energy levels

Ultraviolet (UV)-Visible (Vis) Spectroscopy:

Some organic molecules that show absorption of UV-Vis region of electromagnetic radiation, can be analyzed by UV-Vis spectroscopy. The UV-Vis absorption of electromagnetic radiation is due to electronic transition of the molecules from ground to excited states. Since electronic states are quantized or well separated, the energy absorption by molecule is also quantized. A simple illustration of electronic transition of a molecule due to UV-Vis absorption of radiation is shown in the figure 8 below.

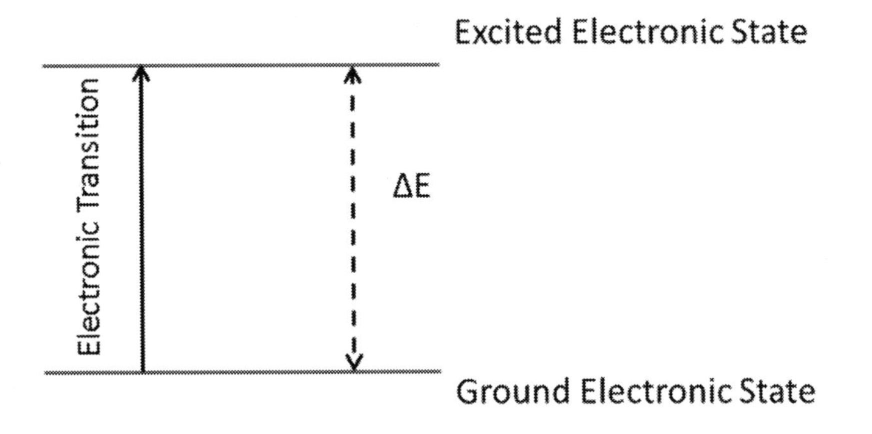

$$\Delta E = E_{excited} - E_{ground} = \text{Quantized or Discrete}$$

Fig.8:

This absorption of electromagnetic radiation results in promotion of an electron from the highest occupied molecular orbital to the lowest unoccupied molecular orbital. For example, in alkane compounds, the highest occupied molecular orbital is σ (sigma) and the lowest unoccupied molecular orbital is σ* and hence, absorption of UV-Vis radiation of alkane compounds is due to σ to σ* electronic transition. Table 7 shows some important but characteristic of highly occupied ground electronic state, lowest unoccupied excited electronic state and electronic transition of some organic compounds.

Table 7: Electronic transition of selected organic compounds

Compounds	Highest Occupied Electronic Ground State	Lowest Unoccupied Electronic Excited State	Electronic Transition
Alkanes	σ	σ^*	$\sigma \rightarrow \sigma^*$
Carbonyl compounds	σ	π^*	$\sigma \rightarrow \pi^*$
	π	π^*	$\pi \rightarrow \pi^*$
	n	π^*	$n \rightarrow \pi^*$
Oxygen & Nitrogen	n	σ^*	$n \rightarrow {}^*$
Alkenes, Alkynes & Azo compounds	π	σ^*	$\pi \rightarrow \sigma^*$

Where

π= Pi bonding orbital; π^* = Pi antibonding orbital; n = non-bonding orbital

σ=sigma bonding orbital & σ^*=sigma antibonding orbital

Below is the order of increasing energy for the electronic transition.
$n \rightarrow \pi^* < n \rightarrow \sigma^* < \pi \rightarrow \pi^* < \sigma \rightarrow \pi^* < \sigma \rightarrow \sigma^*$

It is noted that electron from non-bonding orbital (n) to antibonding pi orbital (π^*) has lowest energy required for the transition from $n \rightarrow \pi^*$ and electron from $\sigma \rightarrow \sigma^*$ has the highest energy electronic transition.

An important selection rule that applies to electronic transition is that change in spin quantum number during electronic transition is not allowed or is forbidden. Thus forbidden transition by selection rule shows lower absorption intensity or peak or not observed. However, allowed transition by selection rule shows intense absorption intensity or peak.

Beer-Lambert Law:

This law is based on experimental results and hence, empirical expression for absorbance for UV-Vis radiation is given by Beer – Lambert Law.

$$A = \log (I_0/I) = \varepsilon Cl$$ for a given λ of UV-Vis radiation.

Where A = Absorbance; I_0 = Intensity Of Incident Radiation;
I = Intensity Of Leaving Radiation; C = Molar Concentration Of Solute;
L = Length Of The Sample Cell (Cm); E = Molar Absorptivity.

Optical Density (OD):

The ratio of I_0/I is called optical density or absorbance.

Chromophores:

Group of atoms that shows a particular electronic transition of UV-Vis absorption is called chromophore. Thus, the hardest part of considering individual electron is eliminated and hence, electronic transition due to group of atoms is considered. The following selection describes characteristic absorption due to electronic transition of chromophores.

(1) Alkanes:

Alkanes have only sigma bonds and hence, only possible electronic transitions are the electrons present in the sigma bonding, viz., $\sigma \rightarrow \sigma^*$. As stated earlier, this transition requires highest energy of any electronic transition and hence, the alkanes absorb UV energy of electromagnetic radiation.

(2) Alcohols, Ethers, Amines and Sulfur Compounds:

Saturated functional compounds of alcohols, ethers, amines and sulfur compounds have electrons in the non-bonding orbitals (n) and hence, the transition involving electron from non-bonding orbital (n) to sigma antibonding orbital (σ^*), which is considered an important transition. This transition requires energy that is lower than that of $\sigma \rightarrow \sigma^*$ transition.

(3) Alkenes and Alkynes:

Unsaturated compounds such as alkenes and alkynes have pi electrons in addition to sigma electrons and hence, the transition of electron from π to π^* is considered as important one. This transition of π to π^* requires energy that is lower than that of σ to σ^* transition.

(4) Carbonyl compounds:

Carbonyl compounds have functional group C=O and hence, this unsaturated carbonyl compounds have non-bonding electrons (n) at oxygen and hence, the characteristic transition as an important is electron from non-bonding orbital (n) to anti-bonding pi orbital (π^*). This transition is dependent on the group of atoms attached to carbonyl group. In addition to n $\rightarrow$ π^* electronic transition, π to π^* transition is also observed in the carbonyl compounds. The substituent that alters the intensity of and position of absorption peak of UV-Vis absorption of carbonyl compounds is called auxochromes. For examples, methyl group, hydroxyl group and amino group are called auxochromes. Due to auxochromes, the following four typical electronic transitions are observed.

a. The shift in the absorption peak to lower energy or longer wavelength by the substituent in carbonyl compounds is called Bathochromic shift. The bathochromic shift is observed for increase in number of conjugated bonds in alkenes. For example, moving from ethylene (CH2=CH2) to Butadiene (CH2=CH-CH=CH2), the bathochromic shift in the $\pi \rightarrow \pi^*$ is observed.
b. The shift in the absorption peak to higher energy or shorter wavelength by the substituent in carbonyl compounds is called Hypochromic shift.
c. If there is an increase in intensity of peak without affecting the peak position by the substituent in the carbonyl compounds, then, this effect is called as Hypochromic effect.
d. If there is a decrease in intensity of peak without affecting the peak position by the substituent in the carbonyl compounds, then, this effect is called as Hypochromic effect.

Mass Spectrometry:

Mass spectrometer consists of three major units. These are

1. Creation of ions from molecules by bombardment of stream of high energy electrons. Ions can be easily accelerated to move fast in an electric field.
2. Application of magnetic or electric field, separation of ions takes place according to mass to charge ratio of ions.
3. These separated ions are finally detected by a device.

Schematic diagram of mass spectrometer is shown in the Figure 9 below.

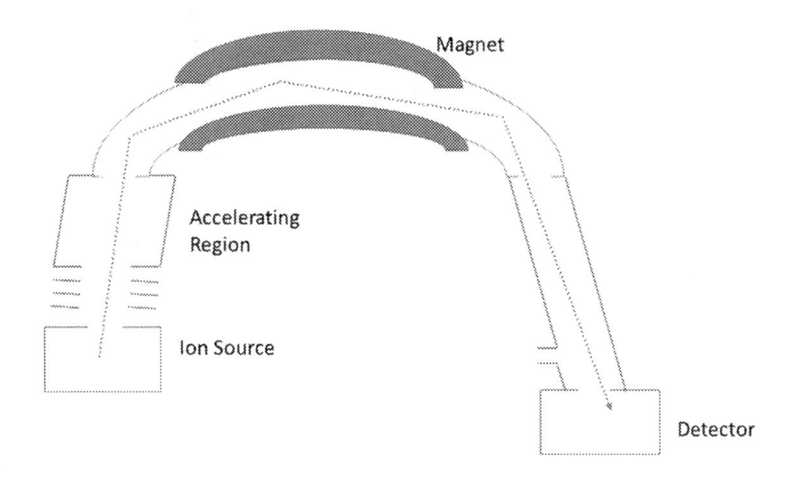

Fig.9: Schematic of Mass Spectrometer

Combination of all the three functions results in mass spectrum, which is a graph of m/z on x-axis and number of intensity of m/z on y-axis. Thus, so called the Base Peak is the highest intense peak that is usually formed from the most abundant ion in the ionization chamber. Due to the application of ionization by the beam of electrons, some of the sample molecules are converted into molecular ion in the ionization chamber. The molecular ion is represented as M+. Thus, m/e relates to molecular ion that is obtained by removal of one electron. The molecular ion is essentially a radical cation since it has both unpaired electron and positive charge as well.

Interestingly, the lifetimes of molecular ions vary according to the following generalized sequence.

Alcohols<branched hydrocarbons < carboxylic acid < ethers < esters < amines < ketones < mercaptans < unbranched hydrocarbons <organic sulfides < alicyclic compounds < conjugated alkanes < aromatic compounds.

This sequence indicates that aromatic compounds are most stable and alcohols are least stable to form molecular ion.

Another interesting rule to state is called Nitrogen Rule. According to the Nitrogen Rule, if a compound having even number of nitrogen atoms (or zero nitrogen atom) is injected to mass spectrometer, its molecular ion that appear in the mass spectrum is at an even mass value. Similarly, the odd number of nitrogen atoms will yield a molecular ion with an odd mass value. The nitrogen rule is illustrated with an example below.

Consider ethylamine, $CH_3CH_2NH_2$ (its odd mass number is 45) and ethylene diamine, NH_2-CH_2-CH_2_NH2 (it's even mass number is 60) and they yield molecular ion of 45 (odd from odd nitrogen containing ethylamine) and molecular ion of 60 (even from even nitrogen containing ethylene diamine) in the mass spectrometer.

Fragmentation Patterns in the Mass Spectrometer:

When a molecule is subjected to high energy electrons in the ionization chamber in the mass spectrometer, the molecule gets fragmented in addition formation of molecular ion due to bombardment of high energy electrons in the ionization chamber. Thus, each molecule has a particular fragment species and molecular ion that are formed in the ionization chamber. The finger print of fragment species and molecular ion of a molecule results in easy identification of molecules by mass spectrometer.

Sometime depending upon the alkyl group attached to benzene ring has resulted in rearrangement which has a typical mass spectrum. For example, McLafferty rearrangement is observed when a propyl group or larger is attached to benzene ring. This can be explained by considering butyl benzene as shown below.

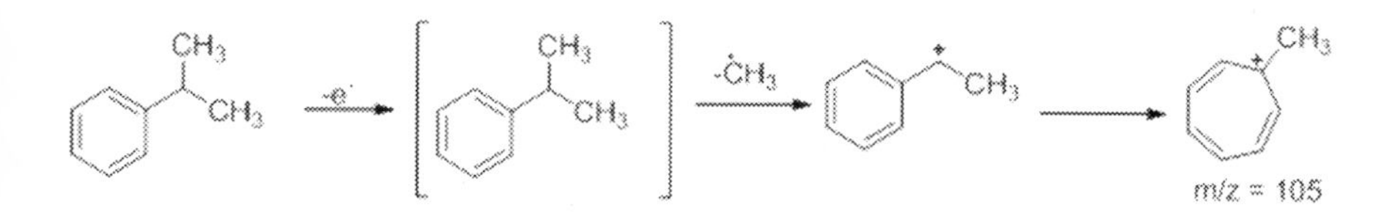

The advancement and innovation in this field is a coupling of gas chromatography with mass spectrometer so that before entering mixture of molecules into mass spectrometer they get separated in the gas chromatography and hence, identification of individual molecule is feasible and facile.

Questions:

1. What is quantum mechanical approach for the observation of vibrational spectra?
2. How does energy level of non-harmonic oscillator differ from harmonic oscillator?
3. What is force constant? Give its application.
4. Explain normal modes of vibration with examples
5. How does Raman Spectroscopy differ from infrared spectroscopy?
6. What is Raman Effect?
7. What are Raman active molecules?
8. What is LASER?
9. What is NMR?
10. What is the effect of magnetic field on nuclear spin states?
11. What is magnetic equivalence?
12. What is Beer-Lambert law?
13. What is mass spectroscopy?

Dyes and Pigments

Objectives:

1. To state nature of visible light
2. To define organic dyes and inorganic pigments
3. To state classification of dyes
4. To elaborate chemistry of azo dyes including synthesis
5. To explain chemistry of C.I. Disperse Orange 25
6. To explain chemistry of C.I. Acid Black I
7. To explain chemistry of C.I. Pigment Yellow 12
8. To explain property of isomerization of azo dyes
9. To explain property of Tautomerism
10. To understand the effect of heat on stability of azo dyes
11. To explain about metal complex azo dyes
12. To elaborate chemistry of carbonyl dyes
13. To explain electrophilic substitution of anthroaquinones
14. To explain electrophilic substitution of carbonyl dyes
15. To state synthesis of Indigoid
16. To show classification of textile dyes
17. To outline reactive textile dyes such as Procion Dye
18. To outline non-reactive textile dyes such as Keratin, self-cleaning function of keratin, acid dyes, cellulose fiber dyes

19. To define inorganic pigments
20. To outline classification of inorganic pigments
21. To outline classification of synthetic inorganic pigments
22. To explain chemistry of TiO_2 pigment
23. To explain chemistry of ZnS
24. To explain chemistry of Iron Oxide Pigment
25. To outline chemistry chromium oxide pigment
26. To explain mixed inorganic pigments
27. To outline very briefly CdS pigment and Ultramarine group pigment

Physics of color deals with fundamentals of visible light, chemistry of color shows how the object or material interacts with visible light and of course, human can observe the color of the material through eyes. Therefore, in this section, physics of visible light cause of color by the interaction of visible light with material and classification of dyes and pigments are the subject of interest.

Physics of Visible Light

Visible light contains wavelength between 380 and 720 nm of electromagnetic spectrum and, our eyes are sensitive to this region of electromagnetic spectrum. Normal white light is usually from this entire visible radiation of wavelength of electromagnetic spectrum. Various colors of visible spectrum with their specific wavelength regions are given in Table 1. These colors are usually extracted by passing entire visible region of electromagnetic spectrum through prism to split and observe the individual color of visible spectrum.

Table 1: Summary of wavelength range corresponding to color and complementary color

Wavelength range, nm	Color	Complementary Color
400 – 435	Violet	Greenish-yellow
435-480	**Blue**	**Yellow**
480 – 490	Greenish – blue	Orange
490 – 500	Bluish –green	Red
500 – 560	**Green**	**Purple**
560 – 580	Yellowish – green	Violet
580 – 595	Yellow	Blue
595 – 605	Orange	Greenish –blue
605 – 750	**Red**	**Bluish -green**

Among the various colors listed in the table 1 the colors, blue, green and red can't be obtained by mixing any of wavelength of visible light, but these three colors in appropriate ways are enough to derive other colors by their mixing. Therefore, the three colors (blue, green and red) are referred to additive primary colors.

Thus, mixing of blue and red produces magenta, blue and green yields cyan, while yellow is obtained by combining red and green. Interestingly, but not surprisingly, combination of all the three primary colors of blue, green and red produces white color due to covering entire visible region. Also, combination of color and its complimentary color produces white color. If an object absorbs a particular color of wavelength, it is the complementary color that is usually observed by human eye. Table 1 does also summarize the complementary colors.

If an object absorbs no light from the visible region of blue, green and red, the appearance of the object is white. Otherwise, if an object absorbs all the radiation from the visible region of blue, green and red, then, appearance of the object is black in color.

As this chapter focuses on organic dyes and inorganic pigments, a brief introduction of these colorants is needed. Thus, color of these colorants is due to electronic transition by absorbing visible radiation and the complementary color of the absorbed wavelength of visible region is their appearance. This phenomenon is completely different from that of fluorescence or phosphorescence. Thus, fluorescence is also due to electronic transition by absorbing visible radiation but absorbed radiation is re-emitted at longer wavelength. Phosphorescence is due to delayed re-emission of fluorescence.

Organic Dyes:

Organic dyes refer to organic compounds with chromophores present in them. Examples include azo compounds and phthalocyanes.

Inorganic Pigments:

Inorganic pigments refer to metal oxides or non-metal oxides with mostly transition metals and rarely lanthanide ions present in them.

Classification of Dyes:

There are two ways by which organic dyes are classified and these are based on

1. Chemical structure of dyes and
2. Use and application method of dyes.

Classification 1: Chemical structure of dyes:

Classification of dyes that are based on chemical structure of dyes refers to the type of chromophores that are responsible for their characteristics colors.

Examples:

Azo dyes [-N=N- chromophore]
Carbonyl dyes [-C=C-]
Phthalocyanines

Classification 2: Applications of dyes

Classification of dyes that are based on use of dyes refer to the dyes that are actually used their properties in the final application of the product.

Examples:
Acid dyes = easy to attach on textiles
Disperse dyes = Easy to impart colors on polyester.
After having seen two major classifications of dyes in particular, individual type of dyes is the subject of interest henceforth. Thus, this starts with chemistry of azo dyes, chemistry of carbonyl dyes, chemistry of phthalocyanine dyes, outline of miscellaneous chemical classes of organic dyes, chemistry of textile dyes, reactive dyes and finally chemistry of inorganic pigments.

Azo dyes: Their chemistry and uses

Azo dyes are class of organic dyes with their common chromophore of azo group, -N=N- linkage, which is attached to SP^2 carbon atoms on both the sides and thus, conjugation of double bond of azo group is extended into adjutant SP^2 carbon. Azo dyes can consist of one azo linkage or two azo linkages or three azo linkages and these azo dyes are called monoazo dyes, diazo and triazo dyes respectively. The adjutant SP^2 carbon is usually benzene rings. The azo dyes can exhibit yellow, orange and red colors. Recent research also has been focused on significant numbers of commercially important blue azo dyes. The azo dyes are cheaper that other class of dyes and hence, they find importance in commercial applications.

Typical examples of azo dyes are shown below.

Mono Azo dyes: Yellow Azo dye

Mono Sulfonated Azo dye: Methyl orange

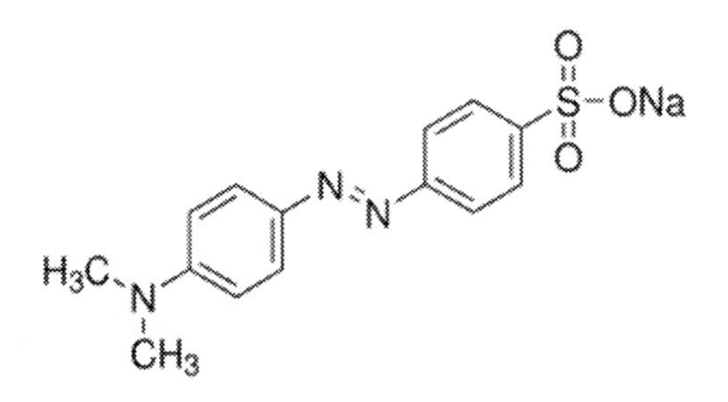

Diazo dye: Sudan 4

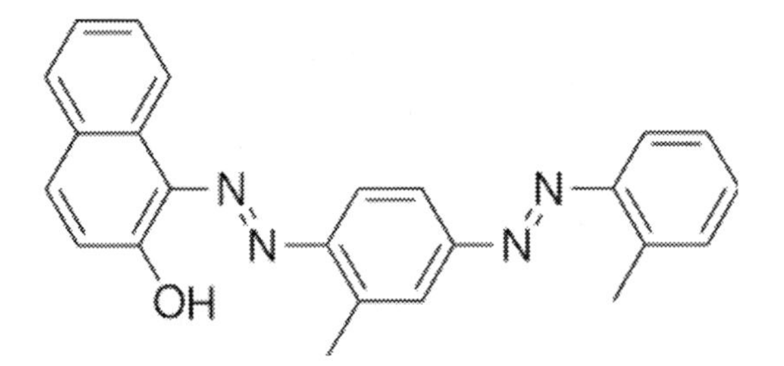

Commercial azo dyes mostly contain hydroxyl group at Orth position to the azo group.

Synthesis of Azo dyes:

Even though several ways azo dyes can be made in the laboratory scale, there are two main steps involved in the individual method of making azo dyes in the large scale. The two steps are

a. Diazotization and
b. Azo coupling.

The chemical reactions involving in the two steps are illustrated below.

a. Diazotization:

In the diazotization step primary amine (for example aniline) is treated with sodium nitrite under controlled acidic condition at relatively low temperature (0 deg to 5 deg C) to form diazonium salt.

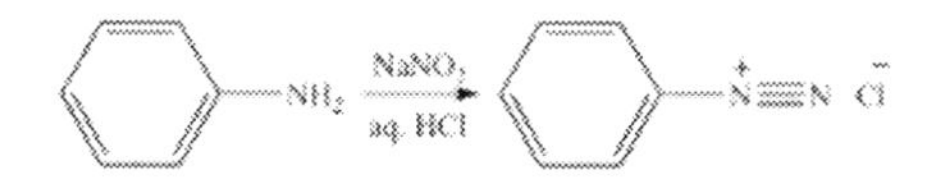

Characteristics of diazotization reaction:

1. Diazotization reaction should be carried out 0 – 5°C
2. Diazotization reaction must be carried out rapidly.
3. Isolation of diazonium salt is not required due to its instability.
4. Diazotization reaction should be carried out under strongly acidic conditions. Thus, at least 2 moles of mineral acid or optimum level of acid are required for every mole of aromatic amines.
5. The diazonium salt/cation can be stabilized by resonance but it decomposes readily with evolution of N_2 gas.
6. The diazonium salt is explosive in the solid state. But, it can be stabilized in the solid state by bulky counter ions other than chloride such as tetrafluoroborate, BF_4^-, tetrachlorozincate, $ZnCl_4^{2-}$ etc.

b. Azo coupling:

In this step after the first step, the relatively unstable diazonium salt that is prepared and stabilized at low temperature is treated with coupling agent such as phenol, primary amine or beta-keto acid derivative to form the azo dye.

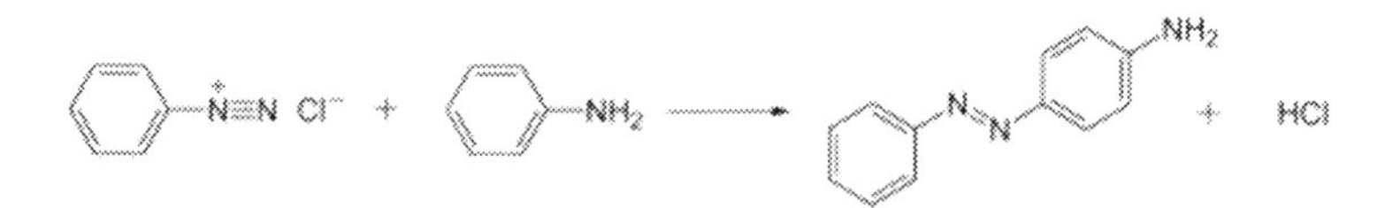

Characteristics of azo coupling reaction:

1. It is an electrophilic substitution reaction wherein diazonium salt is electrophile.
2. Diazonium salt is a weak electrophile and hence, it reacts only with aromatic compounds due to electron-releasing groups present in the aromatic compounds.
3. Azo coupling involves careful control of pH. Thus, optimum pH range is required to have azo coupling reaction in success. When phenol type coupling agents are explored in the azo coupling reaction, slightly basic condition is employed. In the case of aromatic amines, acid condition is preferable.
4. Commonly explored coupling agents are summarized below.

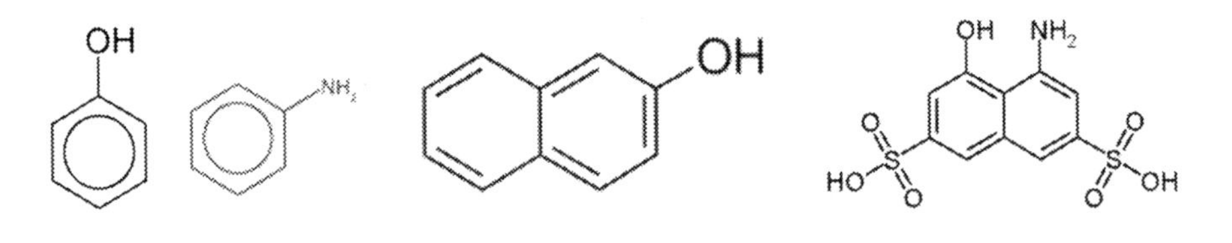

Phenol Aniline Naphthol H-Acid

Selected Examples of azo dyes:

Synthesis of monoazo dyes: C.I. Disperse Orange 25

This monoazo dye is synthesized by the two steps of consecutive reactions.

Diazotization:

The first step is diazotization of 4-nitro aniline using sodium nitrite under acidic condition of aqueous hydrochloric acid at temperature less than 5oC.

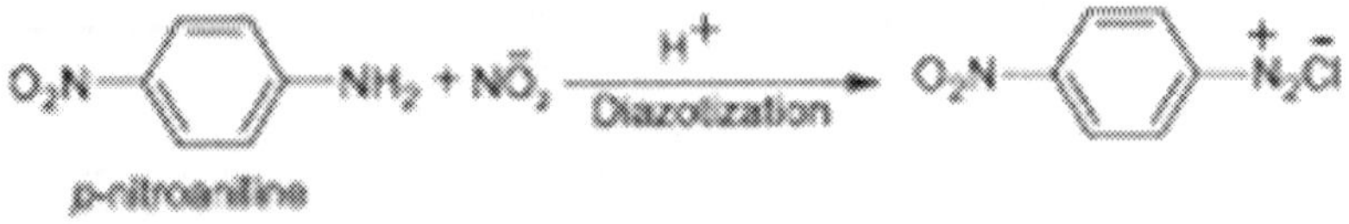

$$O_2N-C_6H_4-NH_2 + NO_2^- \xrightarrow[\text{Diazotization}]{H^+} O_2N-C_6H_4-\overset{+}{N_2}\overset{-}{Cl}$$

p-nitroaniline

Coupling Reaction:

The next and second step is coupling reaction of diazonium cation/salt that is formed from the diazotization reaction in the first step. Thus, diazonium salt is treated with N-ethyl-N-beta-cyanoethylyaniline in weakly acidic conditions to get C.I. Disperse orange 25 as shown below.

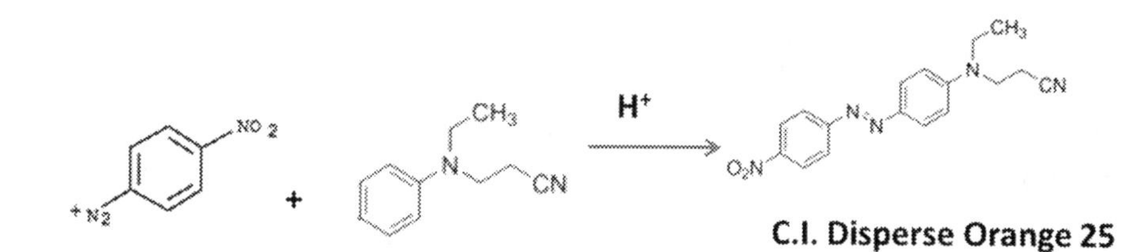

C.I. Disperse Orange 25

2. Synthesis of Diazodyes: C.I. Acid Black 1

Strategies for diazodyes synthesis:

Depending upon the final product of diazodyes, the following strategies are usually followed for achieving diazodyes.

Strategy 1:

Two separate diazotizations with two different reactants are carried out at first.

$$A \xrightarrow{\text{Diazotization}} A^* \text{ (diazonium salt of A)}$$

$$B \xrightarrow{\text{Diazotization}} B^* \text{ (diazonium salt of B)}$$

Secondly, coupling reaction of A* is carried out

$$A^* \xrightarrow{\text{Coupling Reaction}} A^+$$

Then, A+ is treated with B* to get diazodyes as a final product.

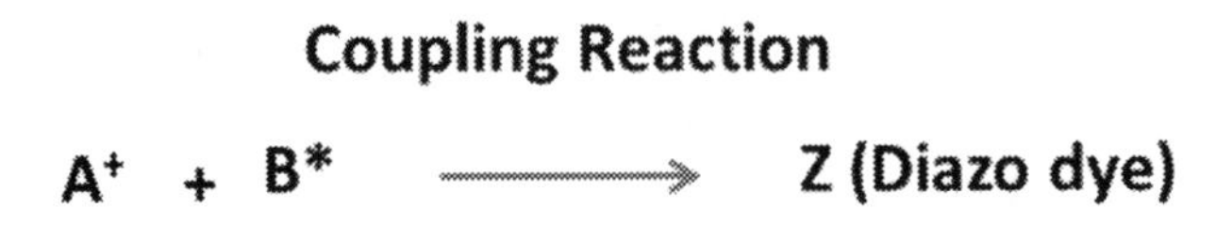

Example for Strategy 1: C.I. Acid Black 1

b. Coupling reactions:

Firstly, diazonium salt of 4-nitroaniline is treated with H-acid (Figure) to succeed first coupling reaction.

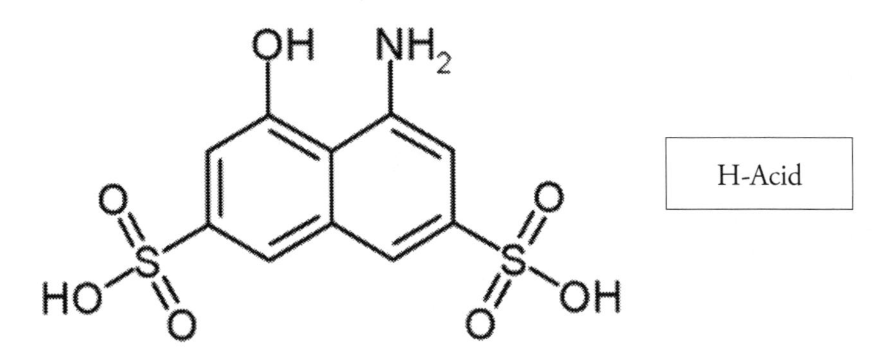

Secondly, the first coupling reaction product is treated with diazonium salt of aniline to get C.I. Acid Black 1 as shown below in a chemical equation.

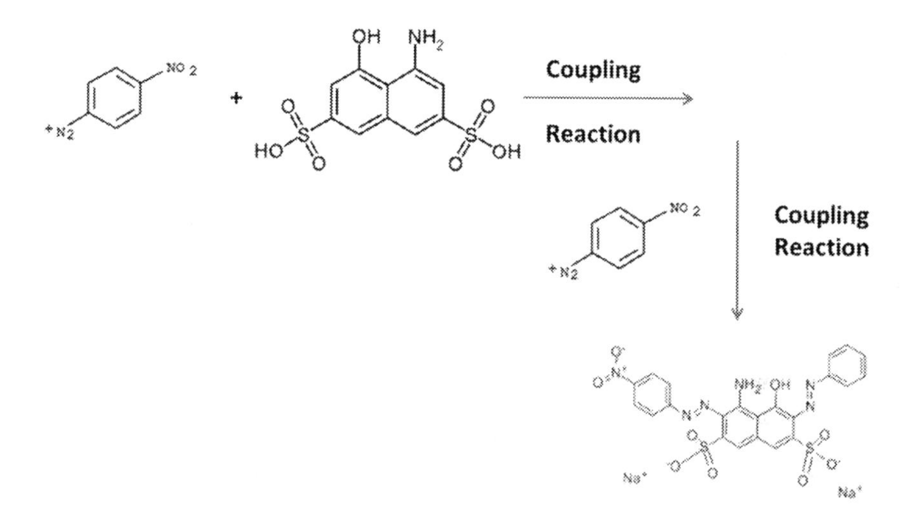

C.I. Acid Black 1

Strategy 2:

In this strategy 2, primary diamine is involved. Thus, first amine group of primary diamine is diazotized followed by another primary diamine is diazotized. Then, the two diazonium salts of single molecule are treated with two coupling components to get diazodyes.

Example: synthesis of C.I. Pigment yellow 12

Bis(diazotization):

The first step is bis(diazotization) reaction. Bis(diazotization) refers to two diazotization reactions on a single molecule. Bis(diazotization) is also called as tetrazotization. Thus, 3,3'-Dichlorobenzidine is bis(diazotized) on both side of amine groups under acidic condition in presence of $NaNO_2$. The chemical reaction involving Bis(diazotization) is written below.

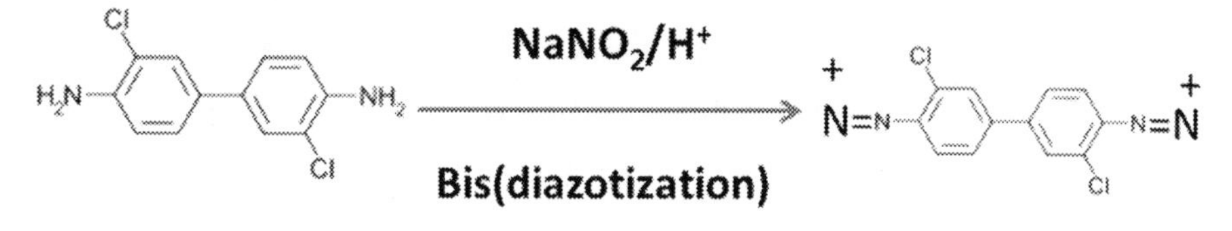

Coupling Reaction:

The next step is coupling reaction. The coupling reagent to get C.I. Pigment Yellow 12 is acetoacetanilide with 1:2 moles ratio. The chemical reaction involving coupling reaction is written below.

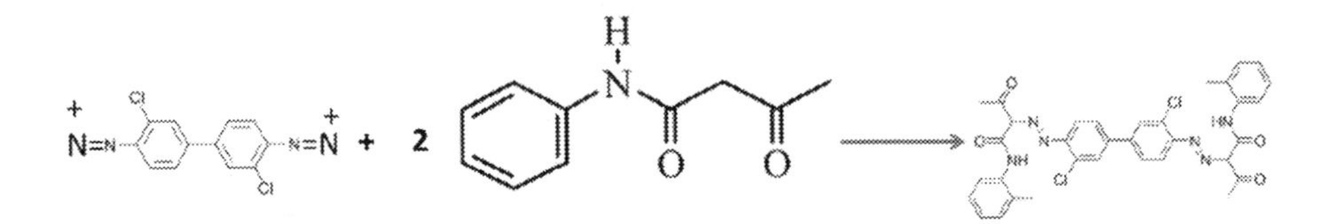

Property of azo dyes:

Isomerism of azo dyes:

There are usually two isomerism observed in azo dyes. One is geometrical isomerism and second is tautomerism.

Geometrical isomerism:

A typical example for geometrical isomerism is with azobenzene is represented below.

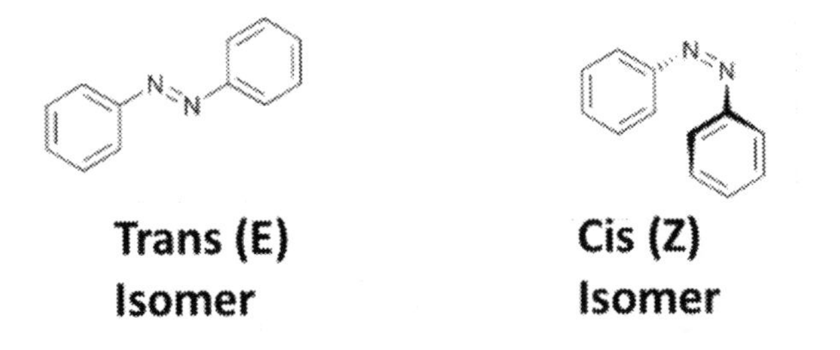

Two possibilities of geometrical isomerism in azo benzene dye

One form is usually stable under synthesis condition, but, the other geometrical isomer can be obtained by photochemical reaction. The two isomers can have different colors. Presence of OH group on ortho position can stabilize trans isomer in particular as shown below.

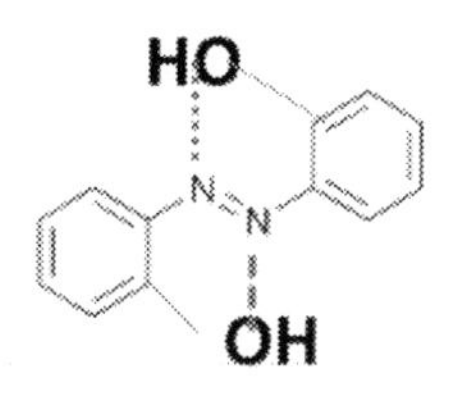

Tautomerism:

When hydroxyl group is attached either ortho or para to azo group an important isomerism of tautomerism is observed. The tautomerism exists in azo dyes only if the hydroxyl group has conjugation with azo group through benzene ring. A typical example for the tautomerism in azo dyes is between hydroxyazo and ketohydrazone tautomerism. Example for tautomerism in azo dye is shown below.

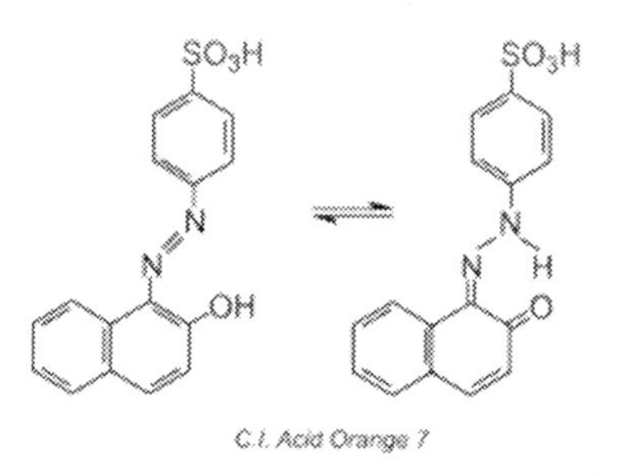

C.I. Acid Orange 7

Both isomers exist in aqueous solution with one of them in higher ratio due to the thermodynamic stability.

Thermal effect on decomposition of diazonium salts:

As stated earlier, diazotization reaction should be very carefully carried out in the temperature range of 0°C to 5°C. The main reason for the low temperature diazotization reaction is due to high temperature effect on decomposition of diazonium salt into thermodynamically stable nitrogen molecule as shown below in the chemical reaction.

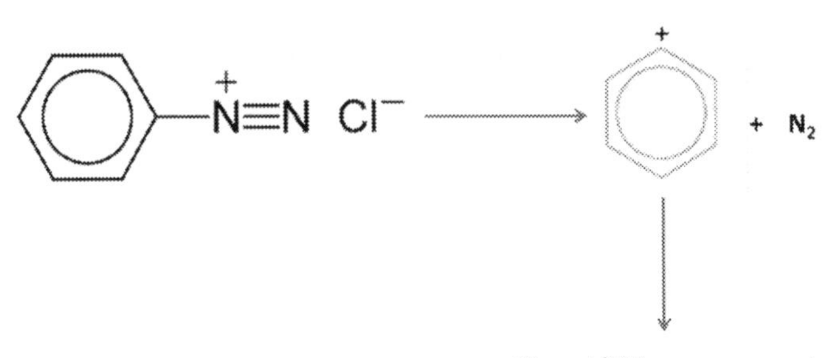

This type of decomposition is exothermic in nature and hence, it can lead to explosive decomposition if care in temperature control is not taken place.

Metal Complex Azo Dyes:

When azo dyes are complexed with carbon metal ions, the process is known as modanting and the metal complex azo dyes are often called as mordant dyes. This type of mordant dyes find very much useful in textile dying due to their high performance of fastness to washing and light when they are compared with that of azo dyes themselves. Mostly, the transition metal ions are explored in the production of metal complex azo dyes.

For examples, copper (II), cobalt (III) and chromium (III) are well-known transition metal ions that are being used for the production of metal complex azo dyes. These metal ions are capable of forming coordination bonds with hydroxyl groups and azo nitrogen of azo dyes. In the case of nitrogen atom of azo group's lone pair of nitrogen atom involves in the formation coordination bonds. The complex azo dye with Cu(II) ion is shown below.

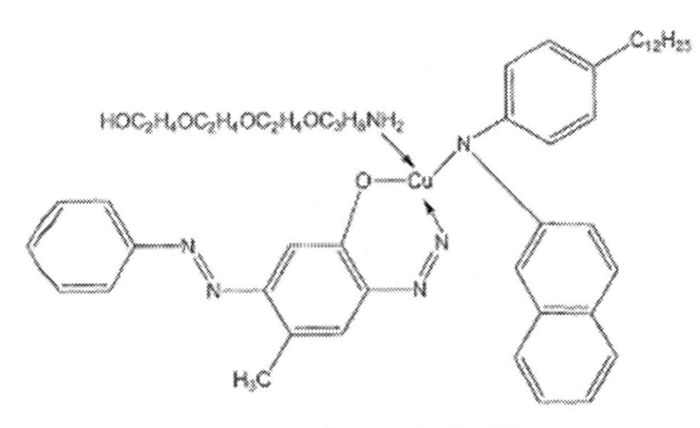

Azo complex with Cu(II)

In this complex, copper (II) has a four coordination environment around it. Three coordination sites arise from azo dyes and fourth site is from monodentate ligand. The geometry around copper (II) in this azo dye is square planar environment.

Cu(II) – azo dye square planar complex thus formed is stabilized by chelate rings. The two types of chelate rings are formed by formation of Cu(II) azo dye square planar complex. One is six membered chelate rings and another is five membered chelate rings.

In the case of cobalt (III) and chromium (III) cations, coordination around these metal ions is six. It is known from the azo dyes that 3 coordination sites can be satisfied by azo dyes and other 3 sites from ligands. Otherwise, two tridentate azo ligands can satisfy six coordination sites for cobalt (III) and chromium (III) cations.

The improved fastness to photo radiation of metal complex azo dye is reduction of electron density in the chromophore.

Reduction of brightness of color of metal complex azo dye is noticed, which is due to broadening of the visible absorption band.

Carbonyl dyes: Their chemistry and uses:

Organic dyes that contain C=O chromophore are called carbonyl dyes. The carbonyl dyes are second class of dyes that are known in the organic dyes chemistry. The carbonyl group (C=O) or chromophore is usually linked to aromatic ring to extend the conjugation of carbonyl dyes. Typical examples for carbonyl dyes are Anthraquinones, indigoids, benzodifuranones, coumaris etc. Figure below shows two examples of structures of carbonyl class of dyes.

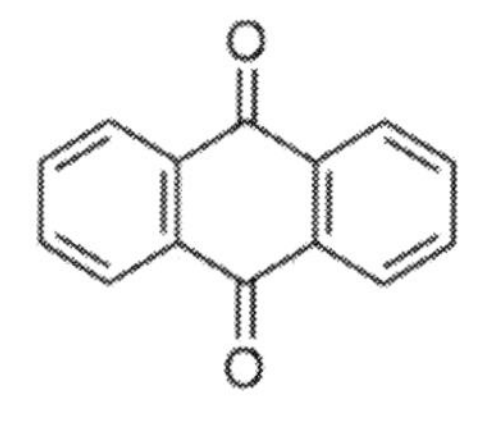

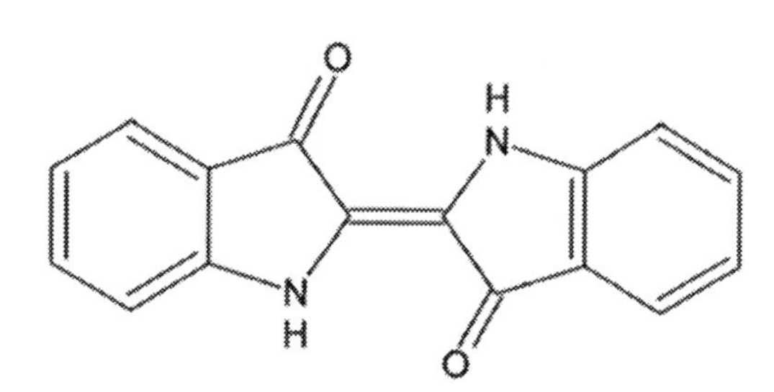

Anthroquinone Indigo

Synthesis of carbonyl dyes:

Unlike common synthesis route available for azo dyes carbonyl dyes have their own synthesis methods depending upon the type of carbonyl dyes. Therefore, selected examples of synthesis of carbonyl dyes are outlined in the following section.

1. Synthesis of Anthraquinones:

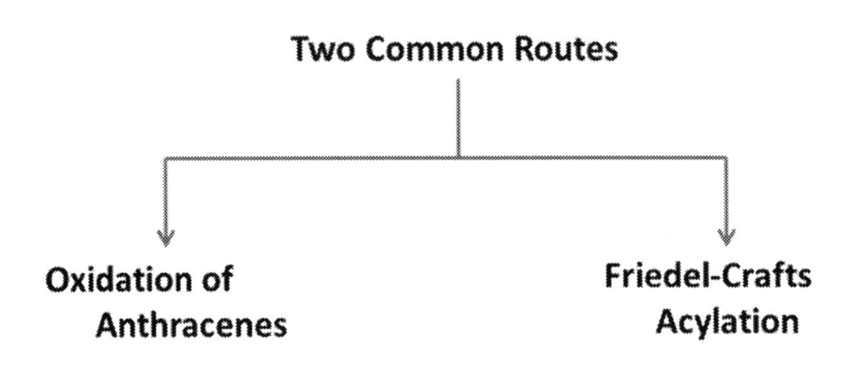

There are two common routes that are widely followed for industrial and practical methods of synthesis of anthroquinones. One of the routes is simple oxidation of anthracenes and another of the routes is Friedel-crafts acylation method.

Oxidation of Anthracene:

Anthracene is a major constituent chemical that is usually obtained from coal tar. Therefore, it is a cheap raw material for the synthesis of anthroquinone. Thus, when anthracene is oxidized to yield in high amount of anthraquinone as shown below.

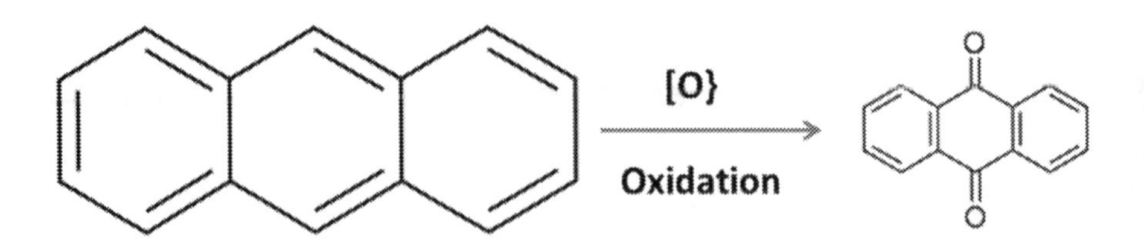

To achieve the oxidation process of anthracene sodium dichromate in acidic conditions is used as oxidant. However, raw materials are not readily available to synthesize substituted anthroquinones.

Friedel-Crafts Acylation:

This route explores phthalic anhydride and benzene as starting materials for the simplest synthesis of anthroquinone. As Friedel-Crafts reaction uses Lewis acid such as anhydrous Aluminum chloride, the reaction leads to formation of 2-benzoyl benzene-1-carboxylic acid. Further acylation by Friedel-Crafts reaction yields anthroquinone due to cyclization under acidic condition. These two steps involve one by one and these two steps are shown below in chemical reactions. This method of synthesis of anthroquinone is versatile due to various substituted benzene availability.

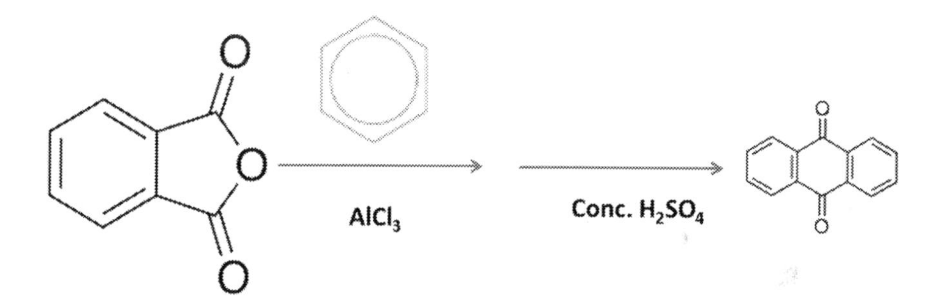

For example, when 4-chloro phenol is treated with phthalic anhydride, quinizarin is obtained by Friedel-Crafts acylation route. Quinizarin is also an anthraquinone. The chemical reaction of Friedel-Crafts acylation is given below.

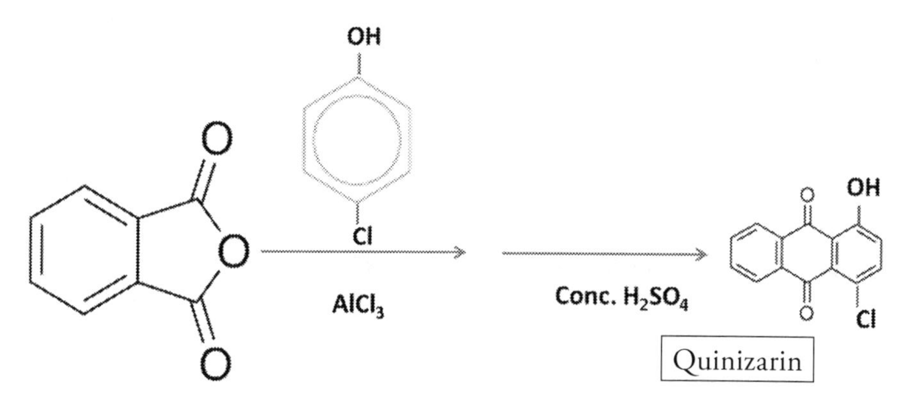

Electrophilic Substitution of Anthraquinones:

Anthraquinones contain two outer benzene rings. Hence, it is expected to undergo electrophilic substitution reactions. But, in principle, electrophilic substitution reactions on anthraquinones are not easily achieved due to the presence of electron withdrawing group, CO in anthroquinone. Therefore, electrons availability in two outer benzene rings are weaker than normal benzene ring. However, in practical, electrophilic substitution reactions can be achieved by moderately vigorous conditions. As representatives, nitration and sulfonation on anthroquinones are given here.

When anthroquinone is treated with concentrated nitric and sulfuric acid, nitration takes place on one of outer benzene rings as shown below.

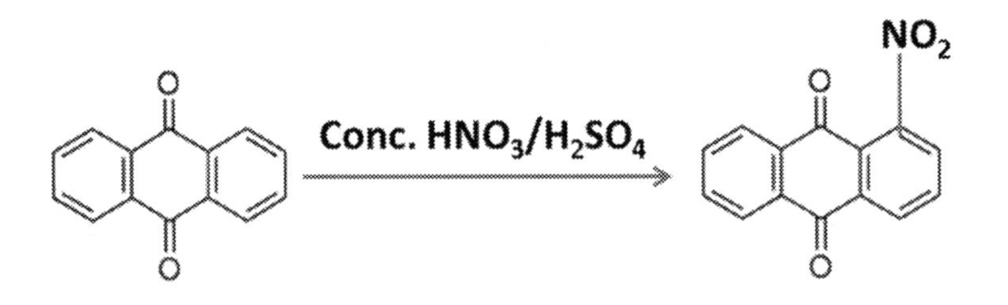

When sulfonation is carried out at elevated temperature it yields 2-sulfonic acid due to steric hindrance at 1-position as shown below.

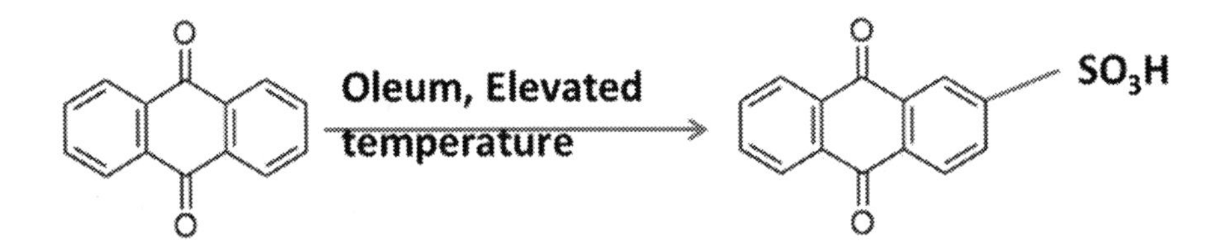

However, when Hg catalyst is used during sulfonation reaction, 1-sulfonic acid is formed as shown below.

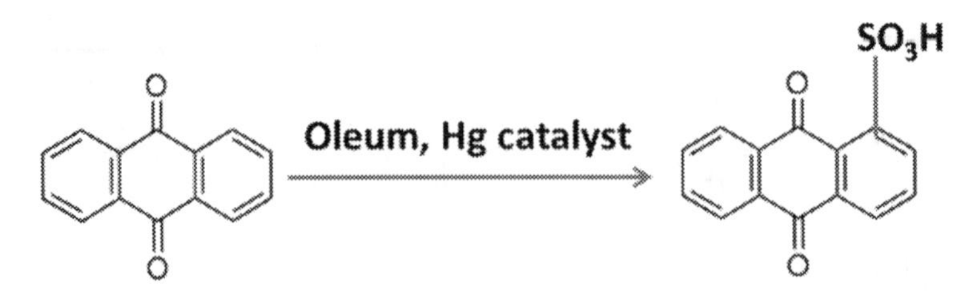

Nucleophilic Substitution Reaction:

The presence of carbonyl groups makes outer benzene rings activated for nucleophilic substitution reaction. Thus, hydroxyl and amino groups can be introduced in the benzene ring via nucleophilic substitution reaction. However, in practice, either boric acid catalyst or transition metal ion catalyst is required.

Synthesis of Indigoid:

Synthesis of indigoid involves more than one step. Thus, fusion reaction of phenyl glycine-O-carboxylic acid using NaOH around 200°C in the absence of air yields indoxyl-2-carboxylic acid. This compound is a reactive compound and it readily decarboxylates and is converted into Indigo in air by oxidation as shown below.

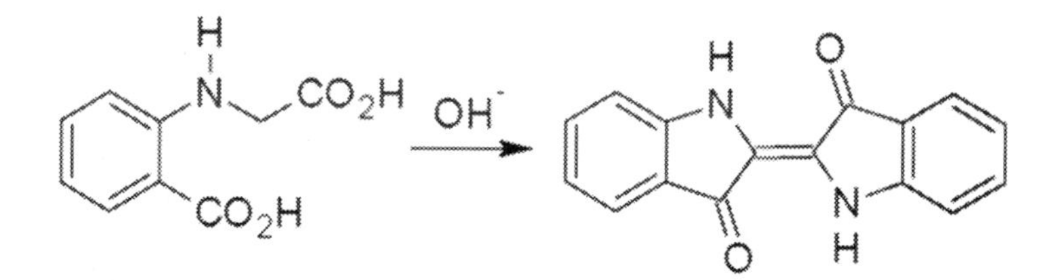

Textile Dyes:

Textile dyes play an important role to impart suitable and appealing colors on a wide variety of products such as clothing, curtains, upholstery and carpets. Therefore, the chemistry of textile dyes with respect to their applications to textile is the subject of this sub-chapter.

Textile dyes are classified into reactive dyes and non-reactive dyes. Description of the textile dyes here is given below in the outline 1.

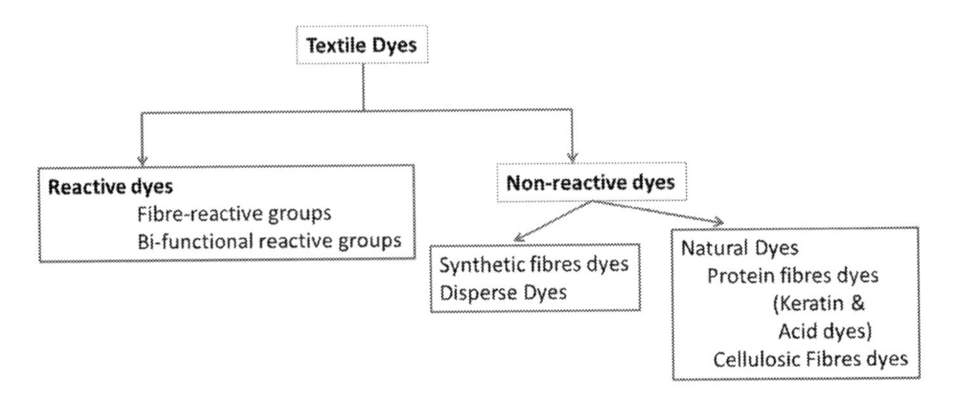

Outline 1: Textile dyes that are covered in this chapter

Reactive Textile Dyes:

A: Fiber Reactive Groups:

Considering cotton fibers, hydroxyl (-OH), amino (-NH_2), thiol (-SH) of cellulose (a polysaccharide) is utilized as nucleophiles for attaching dyes by nucleophilic substitution reaction. Thus, fiber-reactive dye is achieved by aromatic nucleophilic substitution reaction in general and nucleophilic addition to alkenes in particular. Therefore, in this part of sections, how is nucleophilic substitution lead to impart color on fiber using a commercial reactive dues (Procion dyes) illustrated with an example.

Reaction of Procion dyes with cellulose fiber:

Procion dye contains chlorotriazinyl aromatic ring. This dye has major two features to favor nucleophilic substitution through cellulose hydroxyl group. Thus,

1. Nitrogen containing heterocyclic ring has electron withdrawing in nature.
2. Cl atom reduces electron density at the aromatic ring.

The nucleophilic substitution at Procion M dye (also known as Dichlorotriazinyl dye) to impart color on cellulose fiber is outlined with chemical reactions.

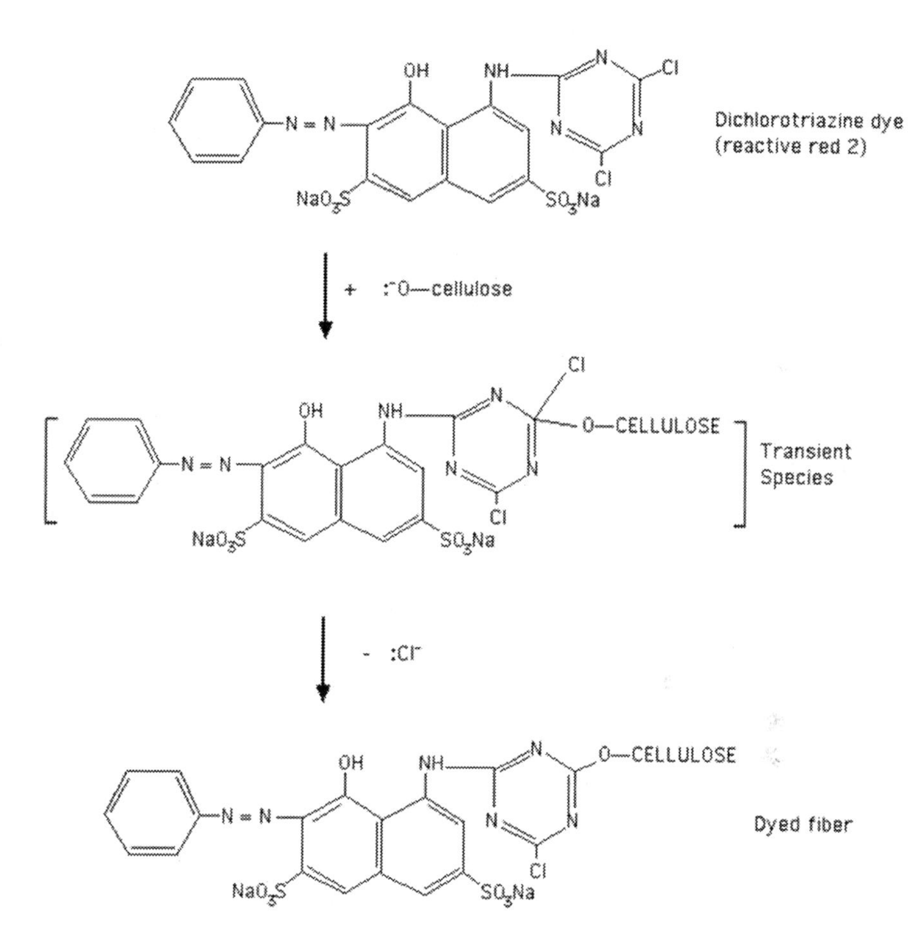

Bi-functional Reactive Textile Dyes:

In the previous section, monofunctional reactive dyes are outlined. But, a major drawback of monofunctional reactive dyes is hydrolysis that takes place in competition with dye-fiber reaction. Therefore, it is highly desirable to minimize hydrolysis of monofunctional reactive dyes to avoid loss of dye molecules without reacting with cellulose. Thus, bi-functional reactive dyes are option to improve dye-fiber reaction.

Non-reactive Textile Dyes:

Keratin:

It is a biological fibrous protein. It is insoluble in solvents due to presence of adjacent peptide bonds. It is a naturally available and is extracted from animal tissues. Typical examples for keratinous protein fibers are wool, silk, hemp and spider. The keratin protein fibers have required physical and processing properties.

Drawback of keratin protein fibers:

Lack of minimum thermal and chemical resistance

Photo degradation in presence of UV in the light.

The basic structure of keratin is given below.

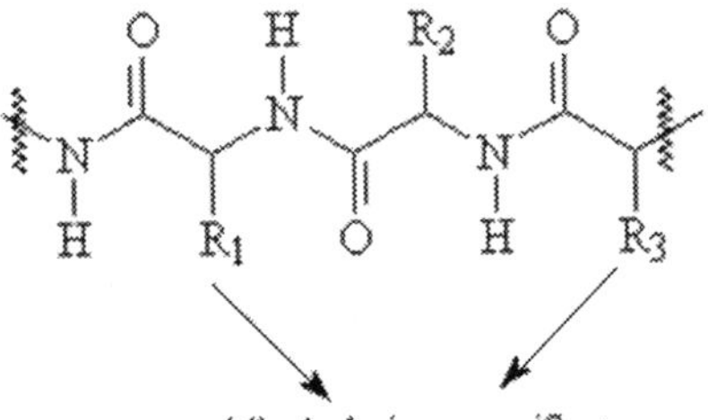

Self-cleaning function of keratin:

Keratin, as a photo catalyst, is able to convert incident light to self-cleaning power to decompose its stains, dirt and harmful microorganisms. This unique property of Keratin can be enhanced by TiO_2 semiconducting Photo catalyst.

Acid Dyes:

Acid dyes are classified to Acid Leveling and acid milling types. Acid leveling dyes have moderate affinity to wool fibers and hence, they do not have a strong force between fiber molecule and dye. The acid milling dyes have a strong affinity to the wool fibers. The acid dyes are soluble in water and the dyes carry negative charge in aqueous solution. The acidic condition is used for dying positive charge of wool. A Typical example for acid dyes for Protein fibers is shown below.

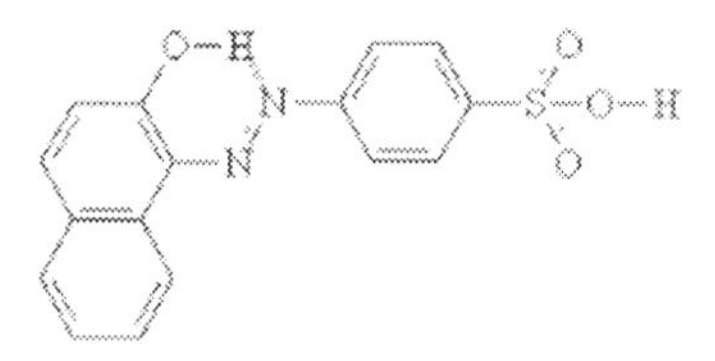

Acid Orange 7.

Cellulose fiber dyes:

Cellulose fibers are yet other natural fibers and they are derived from plant sources. Examples are cotton, ciscose, linen, jute, hemp and flax. Cotton is almost pure cellulose. It is a polysaccharide. Its structure is given below.

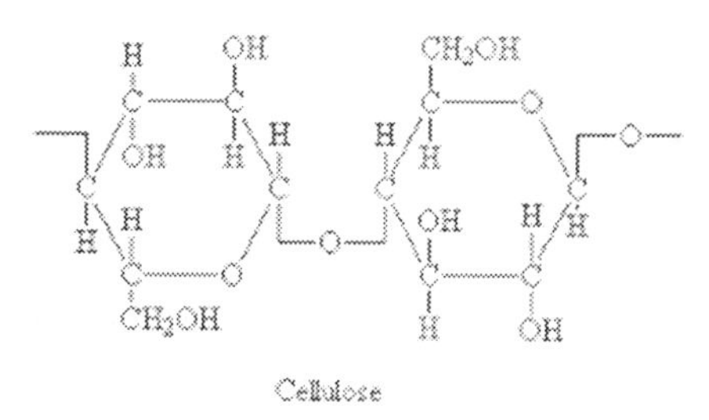

Cellulose

Commonly used dye for cellulose fibers is Direct orange 25. Its structure is shown below.

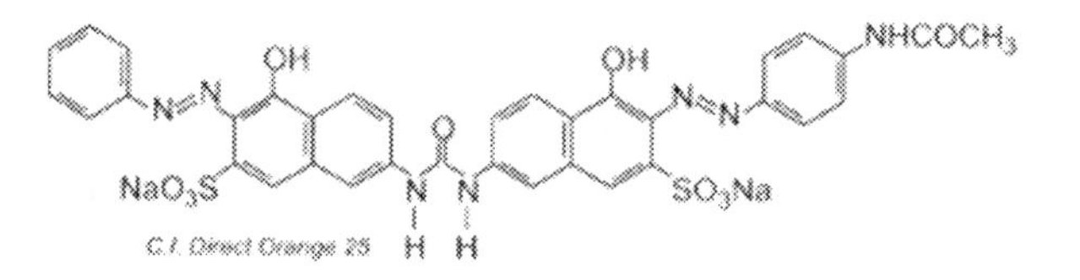

C.I. Direct Orange 25

Synthetic Fiber dyes:

There are three important types of synthetic fibers and these are polyester, polyamides (nylon) and acrylic fibers. Two examples (disperse dyes) for dye used for synthetic fibers are Disperse Blue 165 and Disperse Red 90.

Inorganic Pigments:

Inorganic pigments are solid colorants and they are insoluble in most of the solvents including water. Inorganic pigments are classified into natural inorganic pigments and synthetic inorganic pigments as shown below.

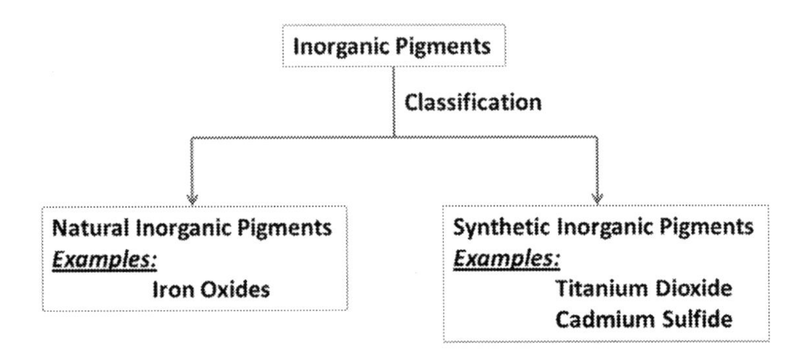

Natural inorganic pigments include iron oxides. Synthetic inorganic pigments are titanium dioxide, chromium oxides and so on.

Thus, the focus of this section is on structural chemistry and properties of selected synthetic inorganic pigments. Optical properties of synthetic inorganic pigments that are responsible for appearance of colors are included as and when required. The synthetic inorganic pigments are classified as white pigments and colored pigments. The synthetic colored pigments are classified as simple oxides/sulfides, mixed oxides and doped oxides as shown below.

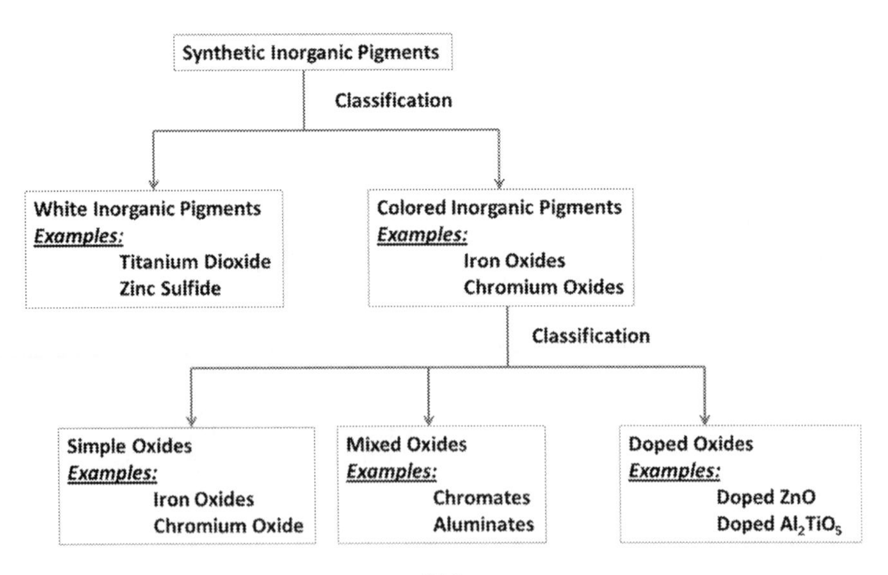

White inorganic Pigments:

(a) Titanium Dioxide, TiO_2:

Titanium dioxide is a well-known and the most widely used inorganic white pigment due to its high opacity. The opacity and whiteness are due to maximum light scattering with minimum light absorption.

Crystal Structure of TiO_2:

There are three types of crystal structures known for titanium dioxide. One is anatase, second is rutile and third is brookite. The anatase and rutile titanium dioxides are being explored in the pigments industry. Anatase is a low-temperature structure and rutile is a high temperature structure of titanium dioxide. Refractive index of rutile titanium dioxide is 2.70, which is higher than that of anatase titanium dioxide, 2.55. As higher the refractive index for rutile titanium dioxide, it is there more opaque. Also, rutile titanium dioxide is more stable and durable than that of anatase titanium dioxide.

Manufacture of Titanium dioxide:

The manufacture of titanium dioxide is achieved from its ore, Ilmenite, $FeTiO_3$. There are two steps involved in the manufacture of titanium dioxide. One is conversion of ilmenite ore into $TiCl_4$ by chlorination of $FeTiO_3$ while reducing iron (II) into iron metal by charcoal. The second step is basic hydrolysis of $TiCl_4$ using liquid ammonia into TiO_2 particles. The chemical equations representing the two steps are shown below.

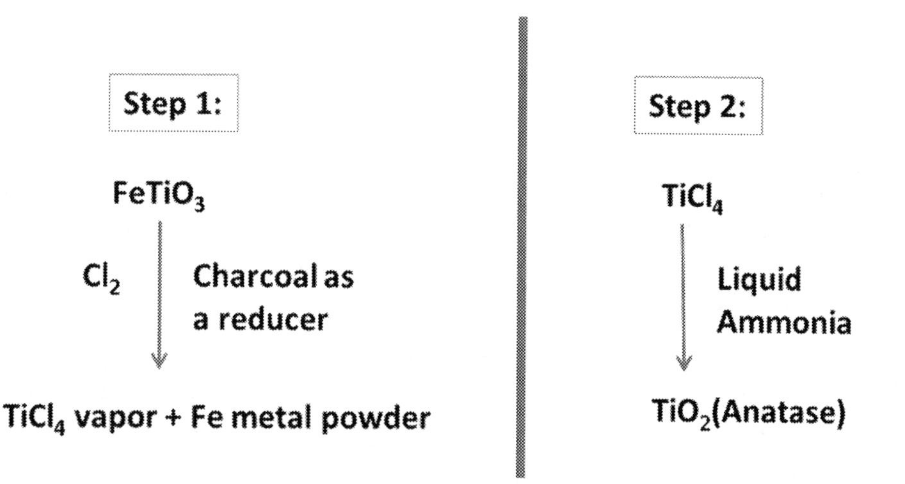

(b) Zinc Sulfide, ZnS White Pigment

Zinc sulfide is another white pigment. It has refractive index of 2.45. The lower refractive index of Zinc Sulfide makes less importance when it is compared with that of titanium dioxide with respect to producing opacity. It is noted that as refractive index is lower it yields more transparent and less opacity, and as refractive index is more, it exhibits less transparent and more opacity.

Other important examples for white pigments are antimony (III) oxide (Sb_2O_3), calcium carbonate ($CaCO_3$) and barium sulfate ($BaSO_4$).

Applications of white pigments:

For example, titanium dioxide (TiO_2) finds wide-spread applications for use in paints, plastics, printing, inks, rubber, paper, synthetic fibers, ceramics and cosmetics.

Synthetic Colored Pigments:

(a) Simple Oxides:

(i) Iron Oxide pigments:

Iron oxide pigments form an important class of simple synthetic colored pigments. Their synthetic methods offer chemical purity and improved control of physical forms. Thus, iron oxide pigments include

Red iron Oxide pigment
Yellow iron oxide pigment
Black iron oxide pigment and
Brown iron oxide pigment

Red Iron Oxide Pigment:

Its main component consists of anhydrous iron (III) oxide, Fe_2O_3. Red iron oxide pigment, Fe_2O_3 is characterized by alpha-crystal modification.

Yellow Iron Oxide Pigment:

Its main component consists of hydrated iron (III) oxides. Yellow iron oxide pigment is represented by iron (III) oxide – hydroxides, FeO(OH). Yellow iron oxide pigment has less stability against heat due to loss of hydration into anhydrous iron (III) oxide.

Black Iron Oxide Pigment:

Its main component consists of non-stoichiometric and mixed iron (II)/ iron (III) oxide. Black iron oxide pigment is represented by Iron (II, III) Oxide, Fe_3O_4. Thus, oxidation states of iron are +2 and +3. These two oxidation states of iron are equally present in the black iron oxide pigments.

Brown Iron Oxide Pigment:

Its main component consists of either the mixed Fe(II)/Fe(III) oxide or the mixture of Fe_2O_3 and FeO(OH).

Properties of iron oxide pigments:

They have excellent durability indicating color of them does not get faded with time.

They have high opacity indicating all of them have high refractive indices

They are low toxicity and

They are low cost to manufacture.

Synthesis of iron oxide pigments:

There are two common routes known for the synthesis of iron oxide pigments. One is from thermal decomposition and another is aqueous precipitation method as shown below.

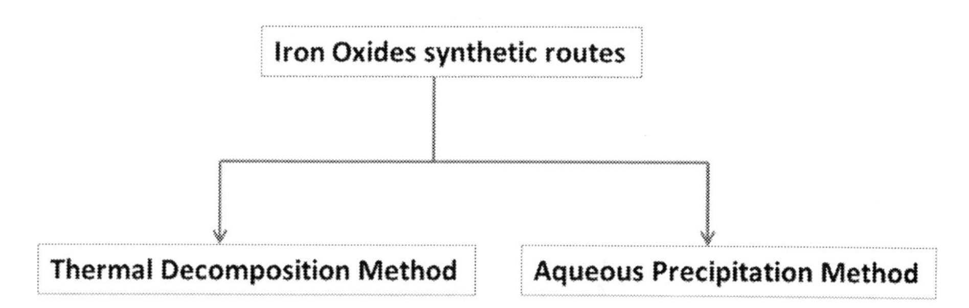

Thermal Decomposition Route:

Thermal decomposition route is explored for the synthesis of Iron (III) oxide with alpha crystal structure. Thus, Iron (II) sulfate seven hydrates is decomposed in air in the temperature between 500 to 750°C to Red Iron

Oxide Pigment. The synthesis of red iron oxide pigment is represented in a chemical equation below.

$$FeSO_4 . 7H_2O \xrightarrow{500^\circ - 750^\circ C} \alpha\text{-}Fe_2O_3 \text{ (Red Iron Oxide Pigment)}$$

Aqueous Precipitation Route:

Aqueous precipitation route is explored to synthesize yellow iron oxide – hydroxide, FeO(OH). Thus, ferrous sulfate is oxidized during hydrolysis to get iron oxide – hydroxide yellow pigment. The chemical reaction in the formation of yellow iron oxide pigment is shown below.

$$4FeSO_4 . 7H_2O + 6H_2O + O_2 \xrightarrow[\text{Precipitation}]{\text{Oxidative Aqueous}} 4FeO(OH) + 4H_2SO_4$$

Yellow Pigment

Pigment Green 17:

Green 17 oxide pigment consists of chromium (III) oxide, and it is represented by Cr_2O_3. It is yet another simple oxide but exhibits yellow color. It is a dull pigment but it offers excellent durability and stability against temperature up to 1000°C.

Preparation of Green 17 Oxide Pigment, Cr_2O_3:

Its simplest method of synthesis is thermal decomposition of ammonium dichromate. The formation of Cr_2O_3 from ammonium dichromate is represented below in a chemical equation.

$$(NH_4)_2Cr_2O_7 \xrightarrow[\text{Decomposition}]{\text{Reductive Thermal}} Cr_2O_3 + N_2 + 4H_2O$$

Green Pigment

Mixed Colored Inorganic Pigments:

Mixed colored inorganic pigments were originally developed for use in ceramics but later they are found useful in plastics. The ceramic pigments are formed due to presence of transition metal ions which impart color on final mixed oxides due to d – d transition of electrons in the visible region. Typical and commercial ceramic pigment is cobalt aluminate blue which is having formula of $CoAl_2O_4$. It is characterized by spinel structure of aluminate. The spinel aluminate based ceramics are called aluminate ceramic pigments. Table 1 below summarizes important metal aluminate spinels and their corresponding shades of colors.

Table 1: Mixed metal aluminate spinel and their colors

Mixed Metal aluminate spinel	**Color**
$CoAl_2O_4$ (Cobalt spinel aluminate)	Blue
$NiAl_2O_4$ (Nickel spinel aluminate)	Cyan
$MnAl_2O_4$ (Manganese spinel aluminate	Blood red to blue, green, brown and colorless

Another important class of ceramic pigments are metal chromites and table 2 below summarizes various metal chromite spinels and their corresponding shade of colors.

Table 2: Mixed metal chromite spinel and color

Metal mixed chromite spinel	**Color**
$MgCr_2O_4$ (Magnesium chromite spinel)	Green
$CaCr_2O_4$ (Calcium chromite spinel)	Green
$MnCr_2O_4$ (Manganese chromite spinel)	Green
$CoCr_2O_4$ (Cobalt chromite spinel)	Bluish-green
$NiCr_2O_4$ (Nickel chromite spinel)	Dark green
$CuCr_2O_4$ (Copper chromite spinel)	Black
$ZnCr_2O_4$ (Zinc chromite spinel)	Green

Manufacture of mixed phase oxide pigments:

Conventional method of manufacture of mixed phase oxide pigments involve high temperature (around 1000°C) solid state reaction of the corresponding amount of individual oxide component. For example, the solid state synthesis of Cobalt mixed aluminate spinel involves reaction between cobalt oxide and aluminum oxide in stoichiometric amount at elevated temperature and prolong annealing. The chemical equation that represents the formation of cobalt aluminate blue pigment is shown below.

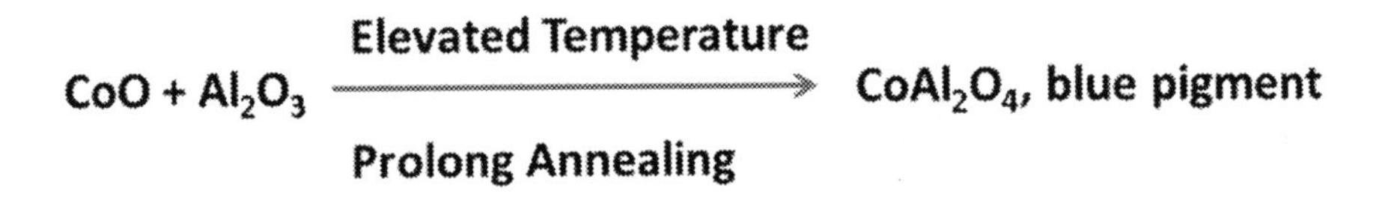

Doped Colored Oxide Pigments:

Doping trace quantity of transition metal ions in white ZnO yields various shades of colors due to occupation of transition metal ion at the Zinc site of ZnO crystal lattice. Such a trace amount of doping of transition metal ions in ceramic ZnO require elevated temperature and prolong annealing. However, aqueous combustion method developed by the author yields not only various shades of colors but also it highly possible for multiple transition metal ions doping in a single matric to have variety of different shades of colors. Table 3 below summarizes thus obtained various transition metal ions doped ZnO ceramic pigments.

Table 3: Various transition metal ions doped ZnO and their corresponding colors

Transition metal ions in ZnO	**Color**
Co^{2+} doped ZnO	Bright Green
Mn^{2+} doped ZnO	Orange
Ni^{2+} doped ZnO	Pale Yellow
Co^{2+} and Mn^{2+} doped ZnO	Brownish Red
Mn^{2+} and Ni^{2+} doped ZnO	Clay
Co^{2+} and Ni^{2+} doped ZnO	Green-rich
Co^{2+}, Mn^{2+} and Ni^{2+} doped ZnO	Dark Brown

Miscellaneous Pigments:

(a) Cadmium Sulfide, CdS based Pigment:

Cadmium sulfide and sulfoselenide [CdS and Cd(S,Se)] form a class of anion solid solution between sulfide, S^{2-} and selenide, Se^{2-} colored pigments. They form range of colors from yellow through orange and red to maroon depending upon the solid solution composition between sulfide and selenide.

Interestingly, when cadmium is replaced by Zn2+ ion, it results in formation of greenish – yellow colored pigment.

Synthesis of Cadmium sulfide based Pigment:

Traditionally, cadmium sulfide pigment is prepared by aqueous precipitation process. Thus, water soluble precursors for cadmium and sulfide are being explored in the synthesis of cadmium sulfide pigment. Cation solid solution between cadmium and zinc ions can be achieved from the corresponding water soluble precursors. In the case of anion solid solution between sulfide and selenide of cadmium pigments, selenide elemental powder is first dissolved in aqueous sulfur solution before subjecting to precipitation reaction.

The precipitation reaction is usually carried out at around room temperature, hence, the resulted cadmium sulfide is in the low temperature beta phase. Hence, the precipitate formed is usually annealed at 600°C which results in formation of high temperature, alpha CdS colored pigment.

(b) Ultramarines group pigments:

It is a natural pigment that is usually obtained from stone. It is usually blue pigment. Its structure consists of three dimensional framework of AlO_4 and SiO_4 tetrahedra units. It can be synthesized by heating the corresponding ingredients to 750 – 800°C over a period of 50 – 100 hrs and the reaction product is cooled in oxidizing atmosphere over several days.

Questions:

1. What is visible light?
2. How does organic dye differ from inorganic pigment?
3. What are the classifications of dyes?
4. What are azo dyes? Give two examples.
5. What is diazotization reaction?
6. What is azo coupling?
7. What are the strategies for diazodyes synthesis?
8. How do you synthesize C.I. Acid Black 1?
9. What is tautomerism?
10. What are carbonyl dyes?
11. Explain Friedel-Crafts acylation?
12. How do you synthesize Indigo?
13. What are bi-functional reactive textile dyes?
14. What is Keratin?
15. State two typical methods for the synthesis of iron oxides.

Printed in the United States
By Bookmasters